Laboratory Handbook for the Fractionation of Natural Extracts

Laboratory Handbook for the Fractionation of Natural Extracts

Peter J. Houghton

and

Amala Raman

*Pharmacognosy Research Laboratories,
Department of Pharmacy, King's College London*

CHAPMAN & HALL

London · Weinheim · New York · Tokyo · Melbourne · Madras

Published by Chapman & Hall, an imprint of Thomson Science, 2–6 Boundary Row, London SE1 8HN, UK

Thomson Science, 2–6 Boundary Row, London SE1 8HN, UK
Thomson Science, 115 Fifth Avenue, New York, NY 10003, USA
Thomson Science, Suite 750, 400 Market Street, Philadelphia, PA 19106, USA
Thomson Science, Pappelallee 3, 69469 Weinheim, Germany

First edition 1998

Thomson Science is a division of International Thomson Publishing I(T)P ®

Typeset in 10/12pt Palatino by Columns Design Ltd, Reading
Printed in Great Britain by St Edmundsbury Press, Bury St Edmunds, Suffolk

ISBN 0 412 74910 6

A catalogue record for this book is available from the British Library
Library of Congress Catalog Card Number: 97–77954

Contents

CHAPTER ONE

Introduction

1.1 THE NEED FOR THIS BOOK – AN ACTIVE EXTRACT, WHAT HAPPENS NEXT?

All over the world at present there is great activity as scientists investigate plants, micro-organisms, marine creatures and many other forms of life for biological activity. There is a desire to find out more about the interactions between one organism and another which can be attributed to the chemical substances present in at least one of the species concerned. The area where activity has been greatest is the effect of extracts from the flowering plants on human physiology and human pathogens, since this is very relevant to the discovery of new drugs for treating diseases of human beings and other mammals. However, it should be noted that other groups of living organisms are now being investigated for the same purposes and regular reports appear in the scientific literature which describe the activity of extracts and compounds isolated from marine organisms, amphibians, fungi and insects.

Other types of chemically based biological interactions are being studied which have implications for other ways in which conditions can be manipulated to serve the human race. Thus, interactions between plants and insects may produce new pesticides or repellents, interactions between two species of flowering plant result in herbicidal compounds and the chemical interactions between species have environmental implications. This interdisciplinary science is sometimes called ecological chemistry and offers many fascinating insights into the complexity of life on our planet.

Many methods have been devised whereby the activity of a compound or extract can be tested scientifically. In the case of possible new drugs, these methods range from the clinical testing of volunteers to cell-free enzyme systems and receptor–ligand binding assays. A similar range of tests, from the 'real-life' situation down to a single important biochemical reaction related to the activity, exists for other areas of investigation.

The experts who devise and carry out such tests are often clinicians, biochemists, pharmacologists, entomologists, marine biologists, plant physiologists and the like, who may have few skills in the area of natural product chemistry. The consequence is that, once having found activity in the extract from a particular organism, there is a desire to find out the compounds that are responsible for the activity but little knowledge of how this can be accomplished. This book is written to give advice and instruction so that further studies can be undertaken to elucidate the nature of the compounds responsible for activity.

There are some excellent texts dealing with the extraction of different types of constituents from plant material (see section 1.5). However, these approach the subject very much from the point of view of the chemical type of the compounds in question and make little or no reference as to whether these are responsible for reputed or observed activity. It is interesting and intellectually challenging to isolate novel chemical compounds, but this book is more concerned with the search for compounds responsible for activity. It is important to note that these may not be the compounds that are isolated most readily or that are of most interest to the natural product chemist. Indeed, one of the drawbacks to the process of bioassay-guided fractionation, upon which this book is based, is that much work can lead to the eventual isolation of a common compound already known to possess the activity under investigation. This has been overcome to some degree by the development of dereplication technologies (section 2.8). These are procedures that eliminate common groups of compounds known to give a positive reaction in a bioassay. The elimination can be performed by selective extraction procedures or by assessing particular physico-chemical characteristics of the components in an extract (e.g. mass spectral profile) and matching them against a suitable database of the same set of characteristics for a large number of known substances in a 'library'. This library is nowadays held as a software database using algorithmic analysis. An example of this for the screening for novel anti-HIV agents is described by Cardellina *et al.* (1993).

1.2 WHY IS IT IMPORTANT TO FIND OUT THE NATURE OF THE ACTIVE COMPONENTS?

It could be argued that the innate curiosity which drives scientific investigation is a sufficient reason for finding out what is causing the particular activity of a plant or its extracts. There are, however, two other major reasons that make such investigations desirable or even obligatory. These are discussed below.

1.2.1 Variation of activity of samples

Any species of living organisms displays a variation within many of its important features due to differences in its genetic characteristics, the environmental conditions under which it is grown and the period in its life history when collection took place. Variation can also occur due to treatment after collection. A common example of such variation is the range of flavour, alcohol content, colour and astringency found in wine. The amount of a particular 'active ingredient' in an organism can therefore vary considerably due to one or more of the above factors and this obviously makes a difference when a particular dose of a crude extract is applied. If the amount present is higher than normal, then toxic effects may be seen. On the contrary, if the amount is considerably lower than normal, the desired effect may not occur.

It is thus obviously important to be able to measure the amount of the 'active' substance or substances present so that some degree of efficacy and safety can be assured. This can only be done once the identity of the actives is known. A large number of very useful drugs and pesticides have been developed using this approach. Prominent examples include the steroids used as anti-inflammatories and for oral contraception, the local anaesthetics based on cocaine, the pyrethroid insecticides, and the muscle relaxants used in surgery which have been derived from the constituents of South American arrow poisons known as curare.

1.2.2 Development of other active substances

The use of natural sources for bioactive substances presents several difficulties, particularly if the material is processed and distributed commercially. One of these is the variation discussed above but there are several other considerations. The regular supply of source material may be uncertain due to political factors such as war and trade embargoes. The supply may diminish or cease due to overharvesting, extinction of the organism in question or environmental disasters such as drought, flood, disease or pest attack. Another factor that influences commercial decisions is the difficulty of obtaining patent cover for a living organism as a source of an active compound or compounds. Most industrialized societies have licensing and registration procedures for biologically active substances of all types which have the aim of ensuring the quality, safety and efficacy of such products before they are officially recognized and recommended for use as medicines, pesticides or similar substances.

In the light of all these problems, it is not surprising that the major activity of commercial organizations is directed towards the discovery and development of individual molecules rather than extracts. The naturally occurring substance may not be ideal due to side-effects, weak activity or a narrow window between effective and toxic dose. A knowledge of its structure is useful, however, since it allows derivatives to be made and can also serve as a template for the design of similar molecules. These compounds may be improvements on the naturally occurring substance.

1.3 GREEN CHEMICALS – THE RENEWED INTEREST IN NATURAL PRODUCTS AS SOURCES OF BIOACTIVE COMPOUNDS

1.3.1 Introduction

The reasons for the revival of interest in biologically active natural compounds are exemplified by changes that have occurred in Western society with regard to pharmaceuticals during the last quarter of the twentieth century.

Most of the earliest pharmaceuticals were plant materials. The obvious effect on human health and activity following the ingestion or application of plant products is known in most societies and the consequent use of plants for treating disease commenced before written history. Similar uses for getting rid of pests have also been known and employed for a very long time. This pragmatic use underlaid the development of modern botany, taxonomy and much chemistry as science progressed. It is noteworthy that all the early investigations into the chemicals found in plants utilized those with medicinal activity. It was found that, in many cases, the activity of the crude plant extract could be reproduced by an equivalent dose of the isolated 'active constituent' and this laid the foundation for the modern scientific presentation of orthodox pharmaceuticals. This active constituent of one molecular type, 'the single chemical entity', could either be isolated from natural sources or synthesized, and was incorporated into a dosage form.

Many naturally occurring compounds were subsequently synthesized and a large number of other related biologically active molecules were introduced which were the products of the genius of synthetic chemists. It should be pointed out, however, that many such compounds were produced based on the structure of natural products.

As the twentieth century has progressed an increasing dichotomy has arisen between the drugs recognized by orthodox medicine and those used in traditional medicine. The former group mainly consists of single chemical entities which have undergone rigorous testing for safety and efficacy before being granted official recognition as medicines. In contrast to this, drugs used in traditional or alternative 'herbal' medicine are usually employed as crude extracts made with water or dilute alcohol. These have little history of scientific testing for efficacy or safety. Those who use such preparations justify their actions on the grounds of tradition and anecdotal evidence.

Two important points should be noted in the context of traditional medicines derived from the flowering plants. One is that they still form the basis for primary health care in 80% of the developing world and the other is the fact that, although their case is judged 'not proven' (often prejudicially) by Western science, some cultures, notably in India and China, have evolved a philosophy of disease and medicine which is very sophisticated and can explain the use of particular plants. It is rather arrogant for Western scientists to claim that a preparation is worthless if it has a long history of seemingly successful use.

For a large part of the twentieth century the use of plant extracts and the potential of the natural world to provide new biologically active compounds has not been taken seriously by many in the pharmaceutical industry. Synthetic molecules were produced and tested in an almost random fashion and this approach has found a modern equivalent in combinatorial chemistry. Recently a more rational approach to the design of synthetic drugs has been made possible because of advances in the knowledge of receptor characteristics and has led to much interesting work.

Just as it seemed that this 'high-tech' approach would result in the production of new single chemical entity drugs for all situations, there was a somewhat unexpected development. Interest and sales in alternative therapies and medicines boomed and provided a fast-growing market in all parts of the industrialized world. Plant-derived extracts, commonly called 'herbal medicine' or 'phytotherapeutic agents' account for the majority of such products.

The past 10 or 15 years have seen several developments which have led to a much greater awareness and interest in natural products in both of these areas. These are briefly discussed below.

1.3.2 Valuable new plant-derived drugs

As well as long-established naturally occurring compounds, such as digoxin, which are drugs of choice in particular conditions, some recently introduced natural products have made significant improvements in the treatment of cancer and other diseases. Compounds such as vincristine, taxol (Paclitaxel®) and etoposide are of outstanding importance in cancer chemotherapy. These molecules have a complicated structure with several chiral centres and, although they have been synthesized, it is more economical to extract them, or molecules from which they can be made easily, from plant material. This point emphasizes the ability of living organisms to synthesize stereospecific complex molecules which may either be used in their own right or act as the basis for chemical or biochemical transformation to other molecules with enhanced or improved bioactivity profiles.

1.3.3 Changes in social attitudes

A change in attitude in Western society has taken place due to the 'Green' movement. This movement has affected the attitude to many facets of life, including drugs, cosmetics and pesticides. The general public has developed a somewhat irrational suspicion of the safety and desirability of synthetic chemicals and has viewed naturally derived substances and extracts as being inherently safer and more desirable. The individualistic philosophy of Western society, accompanied by positive encouragement from government in some cases for financial reasons, has led to an increase in self-medication. Many consumers prefer to treat themselves with phytopharmaceuticals or 'herbal' preparations, and the sales of these are increasing at the rate of about 10% each year in most 'First World' countries.

1.3.4 Multicultural exposure

A related factor which has produced similar results is the exposure to other cultures resulting from the huge increase in international travel, population migrations and worldwide communications. The disenchantment with, and loss of confidence in, many aspects of Western culture and society has facilitated interest and experiment in alternative approaches to life. The treatment and prevention of disease is no exception to this and, since all of the major cultures which have interfaced with the West use plants as medicines, many new products like ginseng from the Far East and guarana from South America now appear in pharmacies and stores dealing with health products.

1.3.5 Ethnopharmacology

The scientific study of the use of plants by a cultural group is called ethnopharmacology and there is now much scientific activity based on this topic since it is realized that much knowledge is being lost due to the breakdown of many traditional societies due to urbanization and the overwhelming Westernization of many aspects of life.

Ethnopharmacological observation in the past has resulted in the discovery of several important drugs such as reserpine and the arrow poison-derived muscle relaxants. New molecules with interesting and novel structures and modes of action are still being isolated from traditional medicines and poisons, and some of these are being developed by industrial and governmental research programmes. Such programmes are often being undertaken in developing countries as an expression of regained pride and confidence in their cultural heritage. 'Local' remedies which are validated by such investigations may also help ease the burden of debt incurred by the import of expensive conventional pharmaceuticals.

1.3.6 Chemical ecology

It is now commonly accepted that a major reason why living organisms produce biologically active compounds is that they form part of the survival strategy in a competitive environment. Complex chemical interactions are often involved in attraction of pollinators and mating partners or the deterrence and repulsion of predators or infecting agents. The production of such compounds is particularly important where the organism is static and, unlike fishes, insects, birds and terrestrial animals, has little ability to move away from a hostile situation. The situation may be exacerbated when the organism is at the limits of its habitat range and is under constant stress. It is therefore possible that organisms in these situations may produce particularly interesting bioactive compounds.

All the above features have resulted in a desire to know more about the chemistry of the flora and fauna of particular localities, especially when it is realized that it is reckoned that scientific study of the chemistry and biological activity has only taken place for less than 10% of flowering plant species. The figures for other forms of life must be much lower.

This need has produced a mobilization of scientists to investigate the biological activity of living organisms and it is likely that most who read this book are pursuing such a line of research. It is unlikely that many such workers have expertise in every area necessary for such investigations. The aim of this book is to help those embarking on bioassay-guided fractionation for the first time whose expertise lies mainly in the area of pharmacology or biochemistry. These skills enable them to devise and use tests designed to show a biological activity in an extract from a plant (or other source). However, such workers are often unsure of the steps to be taken to determine the nature and identity of the compounds responsible for the activity.

1.4 WHAT ARE WE DOING?

1.4.1 Outline of general procedures

All the procedures and exercises mentioned in this book are often used in a sequence of operations which can be described as bioassay-guided fractionation. The steps in this process are outlined below and the bioassay is described as a 'test'. This book is primarily concerned with techniques used in steps 1–6.

- Step 1 Preparation of sample in form suitable for test.
- Step 2 Testing of sample, preferably in a range of concentrations.
- Step 3 If sample gives a positive reaction to test, fractionation of the sample is carried out.
- Step 4 Testing of fractions.
- Step 5 Fractions showing activity are further fractionated, usually by a different method than that first employed.
- Step 6 Testing of each new fraction.

NB Steps 5 and 6 may be repeated several times until one or more pure compounds are obtained which display activity.

- Step 7 Elucidation of the molecular structure of the active molecules.
- Step 8 Determination of the concentration of active compound necessary for a defined biological activity to be observed.
- Step 9 Determination of the amount of active compound present in an active amount of the original extract and comparison with the value found in step 8.
- Step 10 Isolation, synthesis and testing of related compounds.
- Step 11 Investigation of mode of action and metabolism of the active component.

1.4.2 Notes on the steps

Step 1

The method by which samples have been chosen is important. It may be that the organism in question is related to one already shown to have activity and therefore is expected to contain similar active constituents. In this case the extract should be prepared in the same way as that used for the original.

Another common approach to choice of samples for investigation is to consider organisms used by the local indigenous population of an area. In the realm of medicine this is called the ethnopharmacological approach. If this approach is utilized, then it should be ensured that the extracts tested are produced in as similar way as possible to that used traditionally. Note should be taken of the time of collection, the part used, the solvent, time and temperature of extraction and the use of additional substances.

If a variety of samples are to be tested then they should all be extracted the same way prior to testing. The solvent used to make the final dilutions for testing should be chosen carefully to avoid its interfering with the test procedure. The design of the bioassay should incorporate controls to cover this matter.

Step 2

Although a variety of tests could be, and are, deployed in such procedures, the search for active compounds has been greatly aided by the introduction of microassays which permit the use of very small amounts of material and are ideal for testing fractions. In addition a range of concentrations can be used to give quantitative information about the activity of the sample in question.

There is too large a range of tests for this topic to be dealt with in any detail and the reader is directed to books dealing with this topic e.g. *Methods in Plant Biochemistry* (eds P.M. Dey and J.B. Harborne), vol. 6:

Bioactive Natural Products (ed. K. Hostettman). Very often a test has been devised by a researcher for a particular activity and new developments are continually occurring.

In brief, however, tests for activity can be divided into several categories:

Tests on whole organisms

The organisms used can range from micro-organisms such as viruses, yeast, bacteria and fungi to human beings. Different organisms of the same species may react in very different ways to the same amount of test substance and so it is important to use numbers large enough to obtain statistically significant results.

Such considerations have implications in the area of cost and ethics, particularly where mammals and human beings are concerned, and so are not generally appropriate for bioassay-guided fractionation.

Tests on smaller and less complex forms of life are often carried out to determine viability, or conversely, mortality when affected by the substance under investigation. Most classical antimicrobial tests depend on this approach and the use of brine shrimp larvae as a means of assessing toxicity (said to indicate useful antineoplastic cytotoxicity) has also been widely adopted. In some circumstances the aim of the tests may not be so much directed towards killing but prevention of growth or inhibition of some metabolic process.

Tests on isolated tissues

The classical form of this test is the measurement of contractions of a piece of gut muscle or similar tissue immersed in a bath of nutrient solution to which extracts are added. Such tests can be quite sensitive and many assay systems have been developed for a range of tissues. Differences in response between similar tissues necessitate replicate experiments and the relatively large amount of sample material often needed for each test means that this is usually not a preferred method to use for testing fractions. However, it has considerable potential when the tissue is sensitive to small amounts of extract or where a substance is produced which is detectable at low concentrations.

Tests on cultured cells

A major advance in biological testing has been the introduction of mammalian and other cell lines which can be sustained indefinitely in culture. Many of these exhibit very specific receptors and are used to determine receptor binding or inhibition studies for extracts and compounds. Layers of cells can be produced in microtitre plates usually containing 96 wells. Each well requires only a small amount of test extract so that several replicates can be performed and a range of concentrations tested. The amount of a metabolic product of the cell, or of the amount of a substance bound to a receptor, is usually measured by a sensitive method such as luminescence, fluorescence, colour or radioactivity, depending on the reagent used.

Tests on isolated enzymes

The effect of an extract on the conversion of a substrate by a standard amount of enzyme can be studied if the product is detectable in some way. Common methods of detection include visible light absorbance, fluorescence and radioactivity. These tests are often now carried out in microtitre well plates.

Step 3

If the sample does not give a positive reaction in the test system it does not necessarily follow that it is inactive *in vivo*.

One or more of the following might account for the lack of activity:

- The test may not detect the particular mode by which the substance exerts its effect.
- Activity *in vivo* might be due to a metabolic product of the constituents of the mixture rather than the constituents themselves. Such metabolism can be due to the gut flora if the substance is ingested by animals or by enzymes present in the liver or other organs of the test organism.

In addition, if the total extract shows activity in the test but little is shown by any of the derived fractions:

- The active compounds may have decomposed during the fractionation process. Many natural products are sensitive to light, oxygen, water and changes in pH and this should always be borne in mind. The component profile of the fractions (usually characterized by chromatography) should be checked against that of the total extract; the disappearance of some components and the appearance of those not previously detected would indicate that decomposition is occurring.

  Many plants contain compounds which prevent such decomposition, e.g. antioxidants, and these may occur in different aliquots from the biologically active compounds after fractionation. The substances of interest are thus liable to decompose much more readily.
- Observed activity in the total extract is due to synergism between components which are separated as a result of the fractionation process. It is necessary to test combinations of the fractions to see if this is the case.
- The active components may remain in the system used for fractionation. This possibility should be considered particularly if chromatography with a solid stationary phase has been applied.
- The active components may have been lost due to a process such as co-distillation or sublimation when liquid fractions are concentrated.

Step 5

It is unlikely that much further separation will occur if the same method is used. If possible, a separation based on a completely different principle from that used for the first fractionation process should be utilized. Thus, if the first fractionation used a chromatographic column with silica gel as the stationary phase, the adsorption coefficients of the compounds in the mixture (a measure of their polarity) are the major determinant factor in separation. It would be preferable to employ a process such as size exclusion (sometimes called gel filtration), where molecular size differences account for separation, rather than another silica-based process.

It should be noted, however, that very different separations can sometimes be achieved with use of the same stationary phase and a different mobile phase.

Step 7 (see also section 8.2.3)

The structural elucidation of compounds can be very complicated and is outside of the main aim of this book. Some excellent texts are available (e.g. Williams and Fleming , 1991) but in most cases workers unskilled in the process should seek the advice and help of a sympathetic natural product chemist.

It is essential that the isolated product is pure, preferably in a crystalline state. It is often difficult to obtain crystals of natural products because commonly only small amounts are being dealt with. If crystals cannot be obtained, the purity of the compound should be shown by the presence of a single zone or peak when it is examined in a variety of thin-layer or gas or liquid chromatography systems, respectively.

The following characteristics of the compound should be recorded:

1. Its melting point – using a calibrated thermometer.
2. Its chromatographic behaviour and appearance in a variety of separation and detection systems.
3. Its optical rotation.
4. Its UV/visible spectrum from 190 to 750 nm. Compounds with a π electron system (i.e. double bonds) absorb in this region and many natural products contain this feature. The more conjugation present in a molecule, the higher will be the wavelengths where maximum absorbance occurs. The absorbance maxima for particular functional groups vary according to other features of the molecule and thus can provide useful clues to the rest of the molecule.
5. Its infrared (IR) spectrum, preferably using a potassium bromide disc.
6. Its mass spectrum. This is useful for estimation of the molecular mass of the molecule as well as for much other useful information from its breakdown pattern of mass fragments that can help in structural determination.

7. Its proton and carbon-13 nuclear magnetic resonance (NMR) spectra. A wide variety of techniques are now available and a great amount of detailed information can be obtained by skilful application of these sophisticated techniques. Advice and help from a chemist with expertise in these areas is necessary.

Techniques 3, 6 and 7 use very expensive equipment and skilled operators and regular maintenance are also needed. In many situations access to the services provided are restricted so collaboration with specialist groups is necessary.

Step 8

Once an active substance has been characterized and obtained in a pure state then different doses of it should be tested for bioactivity so that a dose–response curve can be constructed. It is very common to determine from this the concentration of substance (preferably expressed in molar terms) that gives a 50% response. This is called the EC_{50}, or if the activity is inhibitory, the IC_{50} (section 8.4).

Step 9

There are several methods by which the concentration of a substance in a mixture such as an extract can be determined. These are discussed in greater detail in section 8.3. The concentration of the substance in the extract may be considerably less than that expected from the EC_{50} calculated for the pure substance. If this is the case, then the presence of other active compounds in the extract is likely and synergy may also be occurring.

Step 10

It is common to find several substances in an extract which are closely related in chemical structure since they all arise from the same biogenetic pathway. These will often give a similar colour reaction when detected on thin-layer chromatography (TLC) to the active compound initially isolated.

Alternatively, analogues of the original active compound may be synthesized (this may well involve collaboration with a synthetic chemist) or derivatives may be made by treating the original compound with various reagents, e.g. to produce esters from acids or alcohols or to methylate oxygen or nitrogen atoms around the molecule.

Step 11

Once an interesting activity has been found then it is often considered necessary to determine its mode of action. This may well have been shown by the particular test used but more biological experiments may be required which are outside of the scope of this book.

It is important to estimate the amount of material that will be needed for such tests and its isolation in a pure state may require a scaled-up fractionation and isolation process.

1.5 BIBLIOGRAPHY AND FURTHER READING

Cardellina, J.H., Munro, M.H.G., Fuller, R.W. *et al.* (1993) A chemical screening strategy for the dereplication and prioritisation of HIV – inhibitory aqueous natural products extracts. *J. Nat. Prod.*, **56**, 1123–9.

CIBA Foundation (1994) *Ethnobotany and the Search for New Drugs*, CIBA Foundation Symposium 185, Wiley, Chichester.

Colegate, S.M. and Molyneux, R.J. (1993) *Bioactive Natural Products*, CRC Press, Boca Raton.

Dey, P.M. and Harborne, J.B. (series eds) *Methods in Plant Biochemistry*, Academic Press, London.

Evans, W.C. (1996) *Trease and Evans' Pharmacognosy*, 14th edn, Saunders, London.

Harborne, J.B. (1984) *Phytochemical Methods*, 2nd edn, Chapman & Hall, London.

Ikan, R. (1991) *Natural Products – A Laboratory Guide*, 2nd edn, Academic Press, London.

Williams, D.H. and Fleming, I. (1991) *Spectroscopic Methods in Organic Chemistry*, 4th edn, McGraw-Hill, London.

CHAPTER TWO

Important concepts

2.1 INTRODUCTION

It is important to understand several general concepts to be able to select, perform and modify fractionation procedures so that the activity is retained and ultimately the molecules responsible can be identified. These concepts are related to the molecular structure of all the components of the fractionation system and the environments to which they are subjected at different stages of the fractionation process.

2.2 POLARITY

Polarity is a term that is important when deciding fractionation procedures. It gives some indication of the distribution of electrical charge across the molecule and is particularly related to the number of π electrons and lone-pair electrons in a molecule. Aromatic rings, carbon–carbon double bonds and carbonyl groups are the most common form of the first group, whereas lone-pair electrons are associated with electronegative atoms such as nitrogen, oxygen, chlorine and other halogens. Thus a measure of the polarity of a molecule can be assessed by looking for the presence and number of these features. The relative size of the less polar region of the molecule is also important; the larger this is, the less polar the compound will be compared to related molecules bearing the same functional groups. Examples illustrating these points are shown in Table 2.1. Common solvents are listed in order of increasing polarity in Table 2.2. The types of compounds extracted by these solvents are discussed in section 3.4.

2.3 DILUTION AND CONCENTRATION FACTORS

Tests for bioactivity should involve known concentrations of the extracts or substances present. Ideally a range of concentrations should be used to demonstrate a dose-related effect but this is often not carried out until some activity at a particular concentration is noted. This initial test concentration varies according to the test system but it is common for it to

Table 2.1 Polarity of common solvent molecules

Compound	*Structure*	*Polarity* (*N.B. compounds are listed in order of* ***increasing*** *polarity)*
Hexane	$CH_3(CH_2)_4CH_3$	Very non-polar – no double bonds or electronegative atoms
Toluene	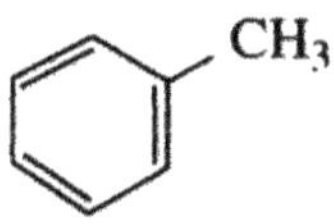	Very slight polarity due to the aromatic ring π electrons
Dichloromethane	CH_2Cl_2	Some polarity due to electronegative Cl atoms
Chloroform	$CHCl_3$	Polarity greater than dichloromethane since 3 instead of 2 Cl atoms
Ethyl acetate	O ‖ $CH_3COC_2H_5$	Appreciable polarity since both π electrons and lone-pair electrons present on O
Butan-l-ol	$CH_3(CH_2)_3OH$	Polar due to free OH group
Water	H_2O	Polar – small molecule and no lipophilic non-polar portion
Acetic acid (ethanoic acid)	O ‖ CH_3COH	More polar than water since ionization to confer negative charge on O can occur

be in the high part of the range required. However, the use, in initial stages, of high and low concentrations, has some advantages, as many beneficial effects at lower concentrations may be masked by general toxicity at higher levels. If an organism is under investigation which has

Table 2.2 Common solvents in order of increasing polarity

Hexane, light petroleum
Benzene, toluene
Diethylether
Chloroform, dichloromethane
Ethyl acetate
Butanone (methyl ethyl ketone)
Acetone
Butanol
Ethanol
Methanol
Water
Acids, bases in aqueous solution

ethnopharmacological use, then the quantities of material, extractant and final extract should approximate to those employed traditionally.

The concentration used should reflect the stage of fractionation and purification. A crude extract from an organism may contain comparatively small amounts (<1% w/v) of an active compound and so extract concentrations 100 or 1000 times greater than those for pure compounds employed as positive controls should be used. Many extracts will show some activity at high concentrations so it is wise to consult the literature to ascertain the minimum level of activity considered as interesting for further investigation.

As the fractionation procedure progresses, the concentration of the actives in relevant fractions will increase, so the concentrations for testing should decrease. This decrease in concentration will depend on the weights of fractions obtained at each stage of the process. It is therefore imperative that each fraction is weighed after it is produced so that its proportion of the total extract can be calculated.

Another important aspect that arises from the results obtained is the translation of the activity into the doses that would be required to use the extract or its isolated constituents commercially. This is not important if the research is purely academic but many investigations are carried out with some regard to eventual utilization and market potential. The concentration of the actives in the organism should be calculated to gain information of the expected yield from a crop or collection. The dosage required for effective treatment is another relevant aspect, particularly when extrapolating the results from testing on animals or in an *in vitro* system as scientific justification or validation of the use in human beings. If the effective dose of an extract or compound is 1 mg/kg in the human and the amount contained in the product used is only 5 mg, then it is unlikely that the product exerts a significant effect. However, it must be stressed that, due to synergy and other effects (section 2.4), the presence of other compounds in an extract may considerably affect the 'effective' dose of a single component present.

2.4 SYNERGY

Synergy occurs when the effect of two or more components applied together to a biological system is more than the sum of the effects when identical amounts of each constituent are used separately.

This may be due to the substances acting in different ways, so that, for example, some rate-limiting step in the process is eased or overcome. Other mechanisms may exist, such as enhanced penetration to the site of action through effects on gut or membrane permeability. Synergy may explain why the activity of an extract or other mixture is greater than might be expected from the amount of the compound, which is thought to be the major 'active' constituent, present.

Synergy may also be one of the explanations for the activities of any of the fractions obtained from a mixture being considerably less than that of the total mixture.

2.5 DETERIORATION, DECOMPOSITION AND STORAGE

The activity of an extract may change over a period of time or after fractionation due to changes in the nature of the chemical compounds present. If those molecules responsible for the activity are transformed to other compounds, the activity of the extract as a whole will change. The more common result of this process is a decrease in activity, but in some cases activity may be enhanced. (The latter case is exploited commercially in the preparation for the market of some products such as vanilla which acquires its characteristic aroma, due to vanillin, only after a period of fermentation.) However, unless such a process is an integral part of the traditional production of the medicine or other product, it is best to minimize the possibility of any such changes occurring.

Changes in the composition of a mixture may be brought about by the action of living organisms or by chemical or enzymatic processes within the mixture itself. The former case is called deterioration and the latter is known as decomposition.

Deterioration occurs most commonly in the biological material which is to be investigated before it is extracted or treated in some other way. Most causes of deterioration are due to microbiological attack, particularly by fungi if plant material is under investigation, but other infestation agents, such as insects, may produce significant changes. Samples consisting of microbial cultures may become contaminated with other micro-organisms. Microbial attack may not only cause some of the constituents present to be changed by oxidation, hydrolysis or polymerization, but the invading micro-organism may produce its own metabolic products which may possess significant biological activity, and therefore confuse the interpretation of bioassay data.

Material to be used should therefore be checked for the presence of infestation visible to the naked eye and may need to be sterilized or extracted quickly with a solvent toxic to micro-organisms. If it cannot be used immediately, it is dried or stored at a temperature well below freezing. Drying is the most common method of preservation commercially and is achieved by leaving the material in warm, dry air. On a small scale, freeze-drying is of increasing importance and is very useful for the preservation of material for laboratory investigation since it makes powdering very easy.

Decomposition is a problem likely to occur at any stage in the fractionation. The major processes likely to occur are oxidation, hydrolysis, isomerization and polymerization. Increases in exposure to light, oxygen, temperature and, in the case of hydrolysis, moisture accelerate such processes. In the crude extract some constituents may prevent the decomposition of others, e.g. antioxidant compounds such as flavonoids may preserve molecules susceptible to oxidation. When these vulnerable compounds are separated by fractionation from naturally occurring preservatives, decomposition is likely to occur. Decomposition may also occur due to interaction with the solvents or other materials used in fractionation procedures. The products formed are called artefacts. Thus the use of aqueous acids or alkali may cause hydrolysis of esters or ester formation and the latter may also be formed in the presence of alcohols if acids are present in the extract. The use of ammonia solution is to be avoided if material is under investigation for the presence of alkaloids, since the nitrogen from ammonia is often incorporated into molecules to yield alkaloid-like compounds that are not naturally present.

Decomposition occurs very rapidly in thin-layer chromatography since the compounds are spread as a thin layer with a large surface area exposed to oxygen. It is therefore vitally important that the bands from a layer separation are removed as soon as possible into an eluting medium (section 7.2).

Breakdown of constituents is often seen as a change in colour. Oxidation of substances, especially phenolic compounds, on a thin-layer plate is usually accompanied by the production of a yellow zone which becomes more intense. Deterioration often accounts for the comparatively small yields of pure active compounds obtained from an extract. It can be minimized by working at as low a temperature as possible, protecting from direct sunlight (or any light) by enclosing the apparatus with aluminium foil, flushing chromatography tanks with nitrogen and by minimizing the amount of time that extracts or fractions are left as thin films on layers or in flasks. Fractions should be stored in closed, robust containers in a refrigerator, and preferably in a freezer at at least −40°C. Glass is a good general-purpose material for containers, although it is breakable. Many plastics are not so useful since they may leach 'filler' compounds into the extract or may absorb components from the extract. This should be borne in mind if plastic caps are used with glass vials, but can be avoided by covering the internal surface of the cap with foil.

2.6 BIOLOGICAL VARIATION

There are two major reasons why biological variation is an important point to consider when extracts are fractionated and tested. The first consideration is the variation in the amount and identity of the chemical constituents found in the material to be examined, which may occur even if the same species is used. The second point of relevance applies where a living system is used in the biological test procedure, since individual organisms or tissues may differ in their response to the same dose of a compound or a particular extract or fraction.

2.6.1 Variation in the chemical profile

Variation in the chemical profile of a source material will probably cause a difference in the activity of extracts when tested biologically. The causes of variability are several. The part of the organism which is used for extraction must be identical, and it is also important to collect the material at the same time of year, and preferably, at the same time of day as previous collections. This applies particularly to plant material, since many studies have showed that the identity and amount of constituents vary considerably according to these two factors. Variation may be due to the existence of distinct phenotypes of a particular species; these are known as 'chemical races'. A particular chemical race will produce much the same profile of constituents wherever it is grown. The total amount of the constituents may vary, but their relative proportions will be much the same.

The second type of variation encountered is that due to external factors such as climate, altitude and soil type, so that the same chemotype will produce a different profile according to its environment. The variation seen in the chemical composition of wine, and hence its aroma, taste and astringency, is due to both the chemical race (the grape variety) and the environmental factors, as seen by the fact that different years of wines from the same plants vary in their taste and other organoleptic properties. Local traditions of collection times and places often provide a clue to the conditions that should prevail for optimum activity to be achieved, and so are worthy of attention.

If multiple collections of material are made, it is important to ensure that all the above factors are standardized as much as possible. Ideally, material should be cultivated, and seed or propagative material from a single known origin used. If a new batch of material is obtained for chemical investigation, its chemical profile should be checked by chromatography against the original material before any work is carried out on it.

2.6.2 Variation in the response of the test system

The biological response to the same dose of any substance will vary between different individuals due to genetic factors, size and several other features, including variation introduced by the analytical method. For this reason it is important that any experiment is designed to take account of this and that the results are subjected to the correct statistical analysis.

To minimize the effects of this variation on the results from a testing procedure several points should be addressed. The biological material used in the test system should be as uniform as possible and all tests should be repeated at least five times. A positive control with a substance of known activity should be included to show that the system is capable of responding in the desired way. Negative controls should also be included to show that any effect shown with the test extract or compound is solely attributable to it and not to components other than the extract. A range of doses of the extract or substance under test should be employed and all responses noted should be quantified in some way.

2.7 AUTHENTICATION OF MATERIAL

It is very important that the identity of any natural material which is intended for investigation should be checked for its authenticity. This is necessary to enable further material to be obtained if the initial tests give interesting results and particularly if compounds with significant activity are isolated. There are many instances where work has been delayed or was unable to be taken further because of confusion or ignorance over the true identity of the samples in question.

The exact geographical source of the material and its time of collection should be recorded. This is especially true for plant material but of less importance for most animals because of their greater ability to move habitat. Representative samples of the organism or its significant parts should be taken and kept in a form suitable for preservation. Fresh plant material is usually dried between absorbent paper and then mounted on a sheet of paper, eventually to be deposited in a herbarium. Microbial organisms and small animal matter can be kept in suitable cultures as live specimens or in preserving fluid as dead matter, and also kept as a reference collection. Identification by an expert on the group of organisms should be made as soon as possible, and ideally should be checked in at least two places against standard reference material in recognized scientific collections, e.g. a herbarium as far as plants are concerned. The criteria used for identification will be those used in mainstream botany, zoology and mycology. These rely on visually observed characteristics.

A representative sample of material from each batch collected is named the 'voucher specimen' and should be given a unique reference number so that it can be traced, if necessary, for further checks on its identity. Any subsequent work published should quote the reference number of the sample under examination.

2.8 DEREPLICATION

A common problem encountered in large-scale screening programmes is that the same, or structurally similar, active compounds can be present in extracts from different species. Since the aim of such programmes is to keep discovering novel lead compounds, it is important that much time and effort is not lost in isolating compounds that are already known to be active in the screen. Thus, a system for 'dereplication' needs to be put in place, which will allow identification of known compounds as early as possible in the process, ideally as soon as a crude extract is found to have activity and before elaborate fractionation steps are undertaken.

Dereplication methods rely on the existence of good databases for active compounds. These contain information such as the source organism, extraction method, chromatographic characteristics (e.g. retention times) of active compounds in defined systems and also the spectral characteristics (usually ultraviolet–visible (UV-VIS) absorption spectra and mass spectra of active compounds). When a new organism is received, it can be checked against the database to ensure that it has not been screened before. Data can also be obtained on any related species that have been screened and found to be active, as these are likely to contain similar chemical components. On confirmation of the biological activity of a crude extract, it can be subjected to chromatographic separation on an analytical scale followed by generation of retention times, UV-VIS spectra and mass spectra of its components. This is done by the use of 'hyphenated techniques' such as GC-MS (gas chromatography–mass spectrometry), liquid chromatography (LC)-diode array detection or LC-MS (liquid chromatography–mass spectrometry) (section 8.2.2), where the spectral method is on-line to the chromatographic separation equipment. The information from the extract is compared against the database, and a decision made as to whether the activity is likely to be due to known or novel substances. In the latter case, the extract is then subjected to further fractionation.

CHAPTER THREE

Methods for extraction and sample clean-up

3.1 INTRODUCTION

Living organisms consist of complex mixtures of chemicals, usually held within cellular structures. In order that these chemicals may be isolated for testing in bioassays, for determining their structure, or both, the initial step clearly involves separating them from the cellular structural material (mostly protein, lipid and polysaccharide) and ideally from the large majority of unrelated substances coexisting in the organism. However, in some cases it may be the structural proteins, lipids or polysaccharides that are desired.

This initial step in separating a part from the whole can be referred to as extraction, and a number of methods exist for this purpose. This chapter will review a range of general methods, methods applicable to specific chemical groups and the factors governing the choice of method. Extraction using solvents which are liquid at room temperature and atmospheric pressure is the most common method. However, other techniques, such as steam distillation, the use of supercritical fluids or press-urized gases, may also be employed.

It should be emphasized that the methods described here will produce extracts which are still relatively complex in nature and would need to be followed by 'clean-up', i.e. fractionation procedures to remove unwanted, unrelated or closely related materials. Clean-up methods are given later in this chapter (section 3.5) and fractionation methods are described in Chapters 4–7. It may also be appropriate to screen the extracts so produced for the presence of specific chemical groups of constituents; chemical tests for this purpose are described in Chapters 8 and 10.

3.2 FACTORS TO BE CONSIDERED IN SELECTING AN EXTRACTION METHOD

A number of extraction methods are described in this chapter and the choice depends on a variety of factors. These factors are discussed in general terms in this section.

3.2.1 Purpose of extraction

Generally, there are four situations:

1. A chemical substance of known identity is to be extracted from an organism. In this case, specific published procedures can be followed, and appropriate modifications made to improve the process or adapt it to the user's requirements.
2. Material is to be examined for the occurrence of a particular chemical group of substances, e.g. alkaloids, flavonoids or saponins, although the exact chemical structures of these compounds, or even their presence, is not known for certain. In such situations, general methods applicable to the chemical group of interest can be obtained from the literature and utilized (section 3.4). These would be followed by appropriate chemical or chromatographic tests for the chemical group (section 8.2).
3. The organism (plant or animal) is used in traditional medicine, and is usually prepared in a certain way, e.g. traditional Chinese medicine (TCM) frequently requires herbs to be boiled in water and the aqueous decoction administered as a medicine. This process must be mimicked as closely as possible if the extract is to be the subject of further biological or chemical scientific studies, particularly if the purpose of these is to validate traditional use.
4. The nature of the substance to be isolated is not predetermined in any way. This situation (encountered particularly in screening programmes) may arise where the aim is to test organisms either selected at random or based on traditional use for the presence of compounds with a specific biological activity. Thus it is necessary to select extraction methods appropriate to the bioassay (section 8.4) and also to try to extract out the widest possible range of chemical types from the organism. This is generally achieved with the use of a range of solvents, but the number of different solvents used will be limited by the scale of the screening programme. Where a small number of organisms is to be examined, a wide range of extracts can be prepared from each sample, whereas in large-scale screening programmes involving thousands of organisms, only one or two extracts (usually of different polarities) are prepared from each.

3.2.2 Scale

For preliminary biological screening, small-scale methods involving a few grams or even milligrams of material often suffice. However, where eventual isolation of active chemicals is proposed, it is necessary to use larger-scale methods capable of handling several kilograms of initial plant or animal material. The apparatus used will therefore differ and it is necessary to be aware that extraction profiles, i.e. the proportion and range of substances extracted, can change when scaling up a process.

3.2.3 Properties of compounds to be extracted

Polarity

Whether the compounds to be isolated are predetermined or not, it is important to note the relationship between the method applied and the properties of the substances extracted. A general principle is 'like dissolves like'. Thus non-polar solvents (section 2.2) will extract out non-polar substances, and polar materials will be extracted out by polar solvents (section 3.4).

Effect of varying the pH

The ionizability of the compounds is another important consideration, as the pH of the extracting solvent can be adjusted to ensure maximum extraction. For example, even non-polar alkaloids can be extracted into polar aqueous acid, as their basic nature ensures salt formation in acid. The salt dissociates into ions in aqueous solutions and the substance dissolves due to hydration of the positively charged, protonated alkaloid and the anion. Solvents at alkaline pH may similarly be used to extract acidic phytochemicals, e.g. fatty acids and phenols. It is important to ensure that the compounds will not break down at the pH values employed, e.g. esters are prone to hydrolysis in alkali and many glycosides lose the sugar moiety in acid.

Thermostability

The solubility of compounds in a solvent increases with increasing temperature and higher temperatures facilitate penetration of the solvent into the cellular structures of the organism to be extracted. However, any advantage gained here will clearly be lost if the compound is unstable at higher temperatures. The formation of artefacts, i.e. new compounds not initially present in the organism under study, is a possibility with many extraction methods. It should be noted that where a traditional method of extraction is followed, these artefacts may actually be responsible for the biological effects observed.

3.2.4 Properties of the solvents to be used

The principle of 'like dissolves like' is, again, applicable here, i.e. the nature of the solvent determines the type of chemicals it is likely to extract from the organism. Other relevant properties of the solvent are the boiling point, flammability, toxicity, reactivity, presence of additives and cost.

Volatility, flammability and boiling point

The boiling point of the solvent, for instance, will determine the ease with which it can be removed from the extract in order to leave the extracted material. The application of heat to remove a solvent may also have deleterious effects on the substances extracted. Thus a low boiling point solvent may be favoured over a higher boiling solvent with the same or similar polarity, e.g. hexane (68°C) over cyclohexane (80.5°C). However, the more volatile a solvent is, the more important it becomes to have adequate containment and safe handling procedures in order to protect the operator and the environment. The hazard potential of the solvent must also be considered. The use of some solvents, such as diethylether, despite their useful extraction properties and low boiling points, tends to be avoided wherever possible due to their extreme flammability (e.g. diethylether, flashpoint −45°C, closed cup). The development of explosive peroxides in ether in the absence of stabilizers is an added consideration.

Toxicity

Another factor influencing the choice of solvent is toxicity to the operator. For example, inhalation of large doses of chloroform or diethyl ether can cause respiratory depression and central anaesthesia. Indeed, due to their lipid solubility and hence ability to enter brain tissue, many solvents are narcotic when inhaled in large doses. Acetonitrile (methylcyanide) and methanol are both considered highly poisonous solvents. Chronic exposure to smaller quantities of solvents may cause more insidious effects such as hepatotoxicity (carbon tetrachloride) and carcinogenicity (benzene). Many solvents cause defatting of the skin, leading to dermatological conditions. Thus less toxic solvents with similar properties must be used wherever possible. The operator must be familiar with the toxic potential of any solvents handled. As a general rule, all forms of external or internal contact with solvents must be minimized through the use of protective clothing, suitable exhaust and disposal facilities and safe working practices.

Reactivity

It is important to be aware that the solvent itself may react chemically with the compounds to be extracted, resulting in the formation of artefacts. The potential chemical reactions occurring in acidic or basic environments have already been mentioned (section 3.2.3). In addition to this, solvents containing carbonyl groups, e.g. acetone and methylethyl ketone, may react with nucleophilic substances in the extract, and the use of methanol or ethanol may result in methylation or ethylation of some functional groups of carboxylic acids. Decomposition products of the solvents can cause damage, e.g. the peroxides which are known to form in ether may result in oxidation or adduct formation with some chemical groups. Low concentrations of stabilizers may be added to solvents, e.g. the antioxidant butylated hydroxytoluene (BHT, 0.0001%) is used in diethylether to limit peroxide formation. Evaporation of large volumes of solvent from extracts can lead to an enrichment or increase in the concentration of these additives. As a result, they are more likely to react with compounds extracted, and may themselves appear in significant amounts as contaminants of dried extracts and isolated compounds. This problem can be overcome by redistilling commercial solvents immediately prior to use in extraction and isolation procedures.

Cost

Where large quantities of solvent are to be used, it may be necessary to use the most economical solvent that fulfils the extraction and safety criteria required. For example, light petroleum (petroleum ether), which is a mixture of alkanes, has very similar extraction properties to hexane. The cost of light petroleum is about half that of a crude hexane fraction (mixture of isomers) and about a twentieth that of purified *n*-hexane.

3.2.5 Use of the extract

The use to which the extract is to be put has a bearing on the extraction method. There are strict laws governing the levels of solvent residues in extracts for food or drug use (oral consumption); the permitted limits vary depending on the toxic potential of the residues. Thus the extraction method must be such as to produce a satisfactory end product that meets the set criteria. Where an extract is to be used for a bioassay, the nature of the test system must be considered. Most biological assays are carried out in aqueous media. Thus the need for water solubility may be a factor which influences the type of extract prepared. An alternative is to use a water miscible general solvent such as dimethylsulphoxide (DMSO) to dissolve non-polar extracts prepared with more lipophilic solvents. However, the sensitivity of the test system to the presence of DMSO must be taken into consideration.

3.2.6 Solvent recycling

The ability to recover the solvent used for extraction may be an important issue for both economical and ecological reasons. Certain methods are more suited to this than others. Where non-azeotropic solvent mixtures are used (e.g. chloroform : methanol 1 : 1) recovery of the solvent in this composition is virtually impossible and separation of the two solvents would require complex distillation procedures. Thus the use of single solvents rather than mixtures may be preferred. Needless to say, where recycling is not possible, adequate care must be taken in disposing of solvents in a way that causes least damage to the environment. Most scientific establishments have standard procedures for the disposal of organic solvents.

3.3 GENERAL EXTRACTION METHODS

The most popular method of extraction is to use a liquid solvent at atmospheric pressure, possibly with the application of heat. Other methods include steam distillation, supercritical fluid extraction and the use of liquified gases under moderate pressure (Phytosol® extraction, section 3.3.4). The choice of method depends on the factors listed above as well as the intrinsic advantages and disadvantages of the procedures.

3.3.1 Solvent extraction

This method involves bringing the material to be extracted (usually in solid form) into contact with the extraction solvent for a period of time, followed by separation of the solution from the solid debris. The equipment used will vary depending on the scale and the need for elevated temperatures. For ease of explanation the material to be extracted will be referred to as plant material, although the principles will apply equally to material of animal or microbial origin. Generally speaking, the material to be extracted will have been pre-dried to avoid the presence of water in the extracts.

A simple method is percolation – where the material to be extracted is packed into a column with a tap at the lower end and an intermediary filter or sinter to prevent escape of the solid material (Figure 3.1) . The tap is opened, and the extraction solvent (at room temperature or above) is poured in at the top and allowed to trickle through the material. In doing so, chemicals are extracted and can be collected in a suitable container. Evaporation of the solvent results in the dry extract. The process can be repeated as many times as necessary to ensure full extraction. The advantage of this method is that it does not require an additional step where the original cellular material (marc) is filtered off from the extract. Disadvantages are that contact between the solid and the solvent may be uneven or limited compared to methods such as extraction under reflux, and the solvent may cool during the percolation process and thus not dissolve some of the components so efficiently.

Infusions are prepared by leaving the plant material to soak in the solvent (generally at room temperature) for a period of time, with or without intermittent shaking, followed by filtration to separate away the

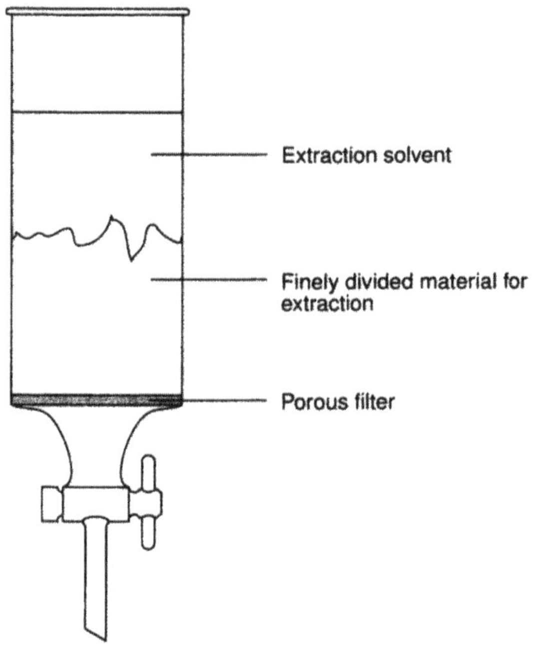

Figure 3.1 Percolation process.

plant debris. If the plant material has settled, then the upper solvent extract can be decanted off, and replaced if necessary with fresh solvent. It is possible to use pre-heated solvents (as in the preparation of a tea!), but these will cool down during the extraction process. The scale of the operation will determine the type of apparatus used – varying from a small flask or test tube to large industrial vessels.

In some forms of extraction, the continuous application of heat may be desirable in order to increase extraction, or to effect a chemical reaction, e.g. hydrolysis of glycosides (section 3.4.5). For extraction with water or aqueous acid or base, heat can be applied directly using a flame. However, it is safer to use a heating mantle as temperature is more easily controlled and there is less likelihood of the glass container cracking. For flammable solvents with boiling points less than 100°C, the container holding the plant material and solvent may be placed over a boiling water bath (heated by the steam) or immersed in a water bath at or below 100°C. Water baths with thermostats are available for this purpose. However, water baths can run dry and, in cases where the set-up is not being monitored continuously, an electrical heating mantle may be preferred. Care should be taken that the solvent does not drip into the mantle. Heating mantles or oil baths can be used in rare cases when temperatures of over 100°C are needed. If only a short period of heating is required, then evaporation of solvent during the extraction process will be minimal. How-

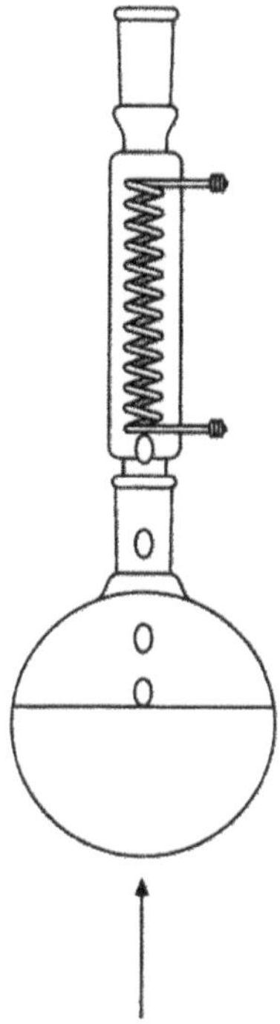

Figure 3.2 Reflux extraction.

ever, when volatile solvents or a prolonged extraction period are used, additional measures have to be taken to prevent loss of the solvent by evaporation during the heating process. These generally involve the use of a reflux condenser, which cools evaporated solvent and allows it to drip back into the extraction vessel (Figure 3.2). An aqueous extract prepared by heating the plant material in water is known as a decoction.

In the method just described, the amount of solvent is fixed and extraction of the chemical constituents from the plant material will cease once saturation is achieved. Thus further extraction would require that the extract is decanted off from the plant material and the extraction repeated with fresh solvent for as many times as required. This method requires a large total volume of solvent and number of manipulations to be performed by the operator. A Soxhlet extractor (Figure 3.3) can be used in order to overcome both of these problems, while still achieving exhaustive extraction by a particular solvent.

In this method, the material to be extracted is placed in a 'thimble' made of cellulose or cloth in a central compartment with a siphoning device and side-arm both connected to a lower compartment. The solvent is placed in a lower compartment and a reflux condenser is attached above the central sample compartment. Note that each component of the

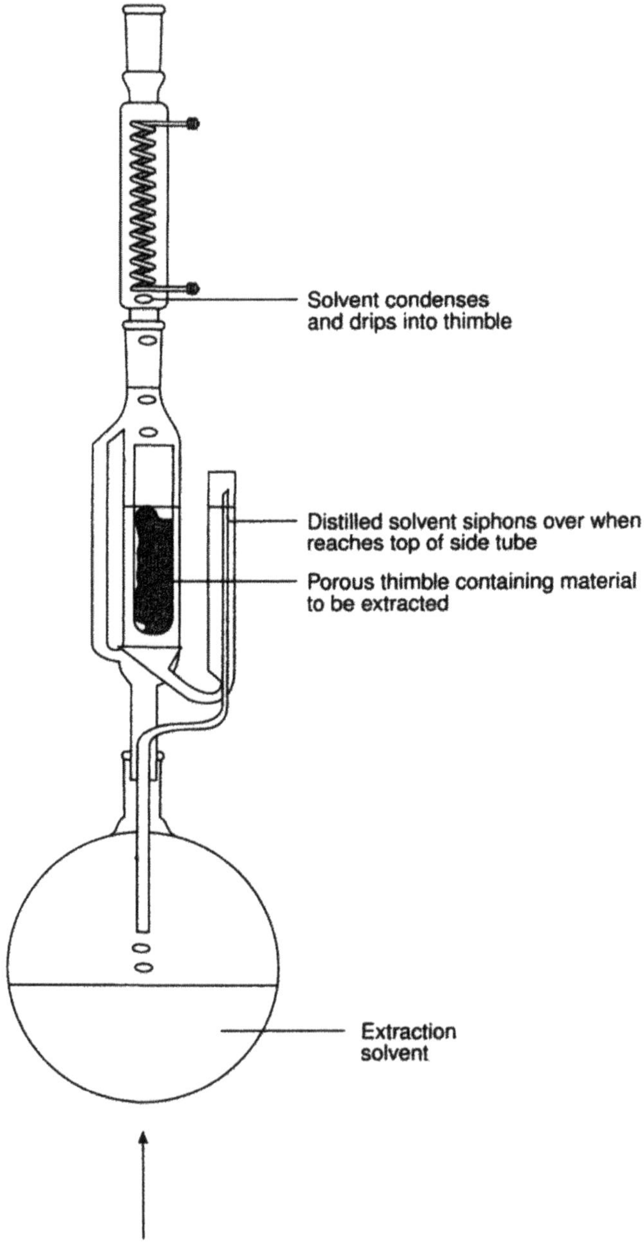

Figure 3.3 Soxhlet extraction.

set up (solvent container, sample compartment and reflux condenser) is a separate item of glassware which is assembled together with the appropriate contents, to make the complete apparatus.

The solvent in the lower container (usually a round-bottomed flask) is heated to boiling, and the vapour passes through the side-arm up into the reflux condenser. Here the vapour liquefies and drips into the thimble containing the material to be extracted. The warm solvent percolates through the material and the wall of the thimble and the extract gradually collects in the central compartment. Once the height of the extract reaches the top of the siphon, the entire liquid in the central compartment flows through this and back into the lower solvent container. The process is then repeated.

In this method, the extract collects in the lower vessel, gradually becoming more and more concentrated. Assuming that no volatile substances are present, the vapour rising from the heated extract is pure solvent vapour and so the liquid dripping into the material from the condenser is essentially pure solvent, though derived from the extract. Thus although a relatively small volume of solvent is needed, the effective volume of solvent used for the extraction is proportional to the time for which the process is allowed to continue.

The Soxhlet process is useful for the exhaustive extraction of plant material with a particular solvent, e.g. for defatting (section 3.5.5) or where 100% yield of a particular component is desired. It is also useful where exhaustive sequential extraction with a series of solvents of increasing polarity is desired, e.g. hexane, chloroform, methanol and water. However, it is necessary to dry the plant material in between changes of solvent to prevent carry-over of traces of the previous solvent into the next one. Varying sizes of Soxhlet apparatus are available to suit the scale of the operation.

This method does have limitations. One major problem is that because the solvent is being recycled, the extract that collects in the lower container is continuously being heated and may suffer thermal degradation reactions. Secondly, the total amount of certain substances extracted will exceed their solubility in that particular solvent. Thus they may precipitate out in the lower container and require a much greater volume of solvent for later dissolution. Thirdly, if the operation is carried out on a large scale, it may not be suitable for use with solvents with relatively high boiling points, such as methanol or water, since the whole apparatus below the condenser needs to be at this temperature for effective movement of the solvent vapour. Finally, unlike the reflux method, this method is limited to extraction with pure solvents or azeotropic mixtures and cannot be used for extraction with any solvent mixture, e.g. hexane : dichloromethane 1 : 1, or acidified or basified solvents, since the vapour would have a different composition to the liquid solvent in the lower flask.

3.3.2 Steam distillation

Steam distillation is a popular method for the extraction of volatile oils (essential oils) from plant material. This can be carried out in a number of ways. One method is to mix the plant material with water and to heat to boiling (distillation with water). The emergent vapours are collected and allowed to condense, and the oil separated from the water. However, if the oils are such that prolonged boiling is to be avoided, then steam from a separate generator can be passed either through plant material which is suspended in water but not boiled (hydrosteam distillation) or directly through the plant material which is laid out on a mesh arrangement between the steam inlet and the condenser (direct steam distillation).

A simple laboratory scale apparatus for hydrosteam distillation is shown in Figure 3.4. Industrial-scale versions are based on the same principles but may employ additional features that allow recycling of condensed water. Once the oil and steam have condensed, the two layers may be separated by physical means (e.g. a separating funnel if done on a small scale); the oil may be the upper or lower layer, depending on its density relative to water. However, where the oil has appreciable solubility in water, better yields will be obtained by solvent extraction of the aqueous layer. For oils with a density equal to or greater than water, an organic solvent, e.g. xylene, which is less dense than water is placed in the collection vessel; the volatile oil dissolves in this upper layer as it condenses. To maximize yields of the oils, precautions must be taken to ensure efficient condensation of the steam and vaporized oil and collection of the condensate in such a way as to prevent loss of the volatile material. However, to avoid risk of explosion, a completely closed system must NOT be used.

Steam distillation relies on the physical principle that, when two immiscible liquids are mixed, each liquid behaves as if it were on its own, and exerts a vapour pressure. The total vapour pressure of the boiling mixture is equal to the sum of the partial pressures, i.e. the pressures exerted by the individual components. Since boiling commences when the total vapour pressure is equal to atmospheric pressure, boiling point is reached at a lower temperature than if each liquid were in its pure state.

Steam distillation requires relatively simple equipment, and no separate filtration step is needed to separate the extracted oil from the plant material. However, it cannot be used where the oil contains hydrolysable components such as esters or those that are easily oxidized or decomposed by heat. For these oils, alternative methods may be used, such as direct extraction from the plant material with fixed oils (enfleurage), solvents, supercritical fluids or Phytosols®.

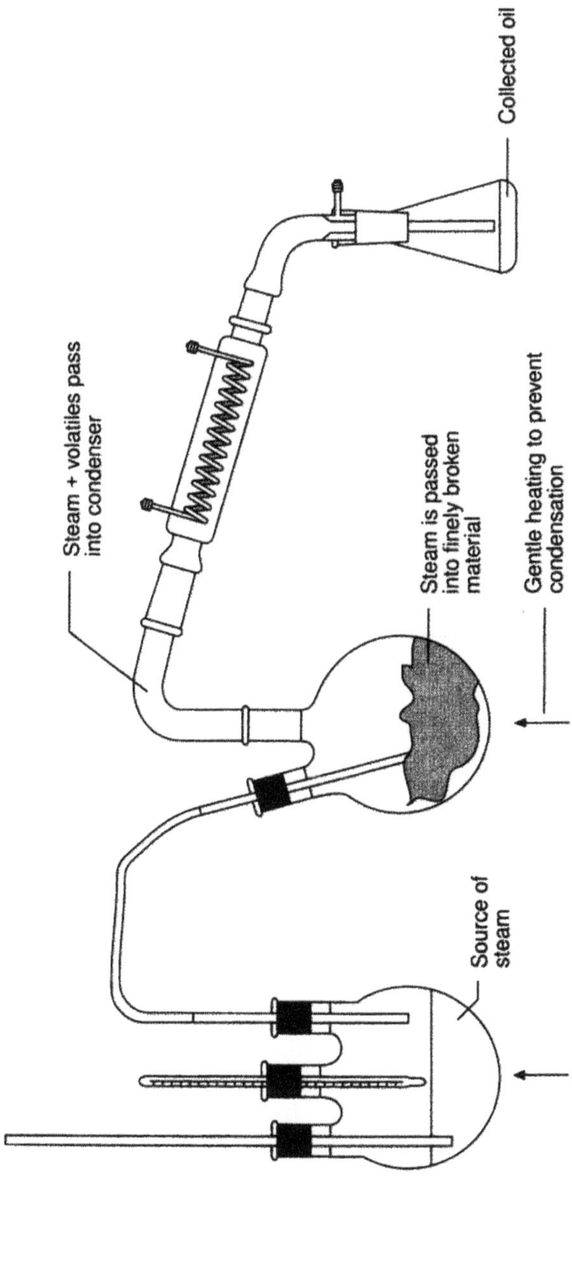

Figure 3.4 Steam distillation.

3.3.3 Supercritical fluid extraction (SFE)

It is well known that alterations in temperature and pressure can effect changes in the physical state of a substance between solid, liquid and gas; these states have marked differences in their densities. However, above a certain temperature and pressure (the critical point), there is no difference in density between the liquid and gaseous forms of the substance and it exists as a fluid that does not condense or evaporate. This is termed a 'supercritical fluid'. The critical temperature and pressure above which supercritical fluids are formed vary with the substance and with its purity. For water the values are 374°C and 220 atmospheres respectively, whereas for carbon dioxide the corresponding figures are 31°C and 74 atmospheres respectively.

Since the early work of Stahl in 1980, it has been known that supercritical fluids can be used to extract natural products. Such fluids have the advantage of low viscosity and are therefore able to penetrate organic material well and achieve good mass transfer. Supercritical carbon dioxide is the most commonly used medium as low temperatures can be employed. This makes it suitable for the extraction of thermolabile compounds. It has other advantages such as non-toxicity, non-flammability and low cost, and can be allowed to evaporate into the atmosphere subsequent to extraction. In terms of its solubilizing effects, supercritical CO_2 resembles the non-polar solvents hexane and benzene, and it has been particularly popular for the extraction of volatile oils from plant material. However, it is possible to improve its affinity for compounds of greater polarity by increasing the fluid density (by adjustments to temperature and pressure) or addition of organic solvents (e.g. methanol, ethanol, dichloromethane) or even water to the supercritical CO_2. These modifiers alter the extraction properties, but it must be remembered that the presence of these compounds will alter the critical temperature and pressure and will also necessitate modifications to the procedures for removing extraction fluid at the end of the process. Selective extraction of a particular component or group is often achievable.

One disadvantage of SFE is that the apparatus required is of a higher level of sophistication than for solvent extraction or steam distillation. The basic instrumentation consists of a CO_2 source (gas cylinder), and a pump to attain the required pressure and deliver the pressurized fluid to the extraction chamber, which is in an oven to allow it to be held at the required temperature. On–off valves are required at the inlet and outlets of the extraction chamber. The fluid leaving the sample chamber can either be led off into an on-line analytical instrument (e.g. a gas chromatograph) or be collected, depressurized and the fluid allowed to evaporate, leaving the extracted material as a residue. Solid samples for extraction need to be finely divided, and liquid or wet samples may be mixed with inert adsorbents (e.g. Celite) or drying agents (e.g.magnesium sulphate) before placing in the chamber.

SFE has been utilized for a wide range of chemical groups and the interested reader is referred to the book by King and Bot (1993) and review by Modey, Mulholland and Raynor (1996). Industrial examples of the use of SFE include the decaffeination of coffee and isolation of bitter principles from hops. Supercritical fluids may also be used as mobile phases in chromatography (supercritical fluid chromatography, SFC).

3.3.4 The Advanced Phytonics process

A patented method called the 'Advanced Phytonics' process has been developed recently for the extraction of volatile oils. It is said to offer better products, in higher yields and at lower costs of production than other processes such as steam distillation, solvent extraction and SFE (Wilde, 1994). In this case, the extracting fluids are named 'Phytosols®', which consist of 1,1,1,2-tetrafluoroethane with or without modifiers (Table 3.1). These substances are gases at room temperature and atmospheric pressure, but are applied to the sample to be extracted under modest pressure (4 bar and over) under which conditions they liquefy. Due to their low viscosity, they have good penetration properties into the material to be extracted.

The Advanced Phytonics method can be applied on a small scale using a hand-held kit available from the Advanced Phytonics company (see end of this chapter for the address). This is designed for use at ambient temperature at an operating pressure of 4–10 bar. The sample chamber is filled with the material to be extracted and an aerosol can containing the pressurized Phytosol® applied. The system is kept under pressure in order to maintain the Phytosol® in liquid form. The extract can then be led into a collection vessel (still in a closed system), the collection vessel disconnected and the Phytosol® allowed to vaporize, leaving the extracted material. Attachments are available that allow the vaporizing Phytosol® to be collected and recycled. Equipment is also available for application of the technique on an industrial scale (Figure 3.5). In this case, higher temperatures and pressures may be applied and solvent is always collected and recycled into the extraction chamber or recompressed for future use.

Some of the advantages of the Phytosol® technique (as compared to SFE) are the lower pressures required as well as the ability to use it at

Table 3.1 Composition of Phytosols® used in the Advanced Phytonics process

Phytosol	*Composition*
A	1,1,1,2-tetrafluoroethane
B	1,1,1,2-tetrafluoroethane + butane/isobutane
D	1,1,1,2-tetrafluoroethane + dimethylether

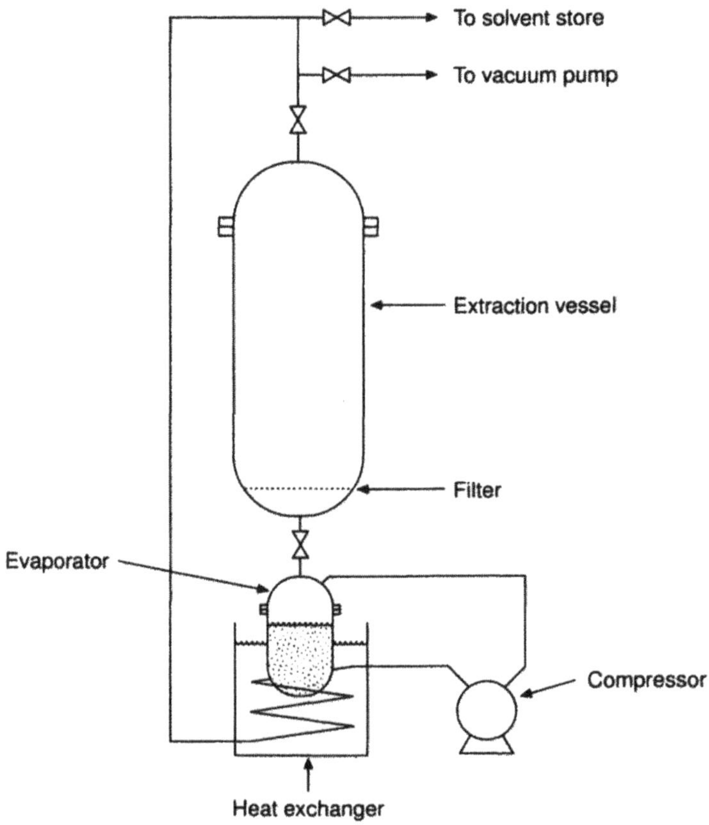

Figure 3.5 Phytosol® extractor (industrial scale).

room temperature. The equipment required is relatively simple in the case of the hand-held kit. Although the three Phytosols® have some polarity differences, they are essentially designed for the extraction of non-polar materials. Their usefulness for polar substances has not been demonstrated.

3.3.5 Miscellaneous methods

Enfleurage

This is a method frequently used in the fragrance industry to obtain volatile oils from flowers. It is a form of solvent extraction where fixed oils are used as the solvent. The oil is spread in a thin layer over a glass plate and the flowers to be extracted are placed on the oiled plates. The volatile (essential) oils in the flowers gradually dissolve in the fixed oil layer. The exhausted flowers can then be removed mechanically or by hand. This method avoids the use of heat which may damage the fragrance oils, but suffers the disadvantage that the volatile oils are not obtained in pure form.

Expression

Expression can be used to obtain fixed oils from plant material. This involves applying pressure to the material, which disrupts the cellular structure and allows the oil to flow out of the material. This method is frequently used for industrial production of vegetable oils such as soya oil, sunflower oil and olive oil. Heat may be applied to make the oils less viscous, but in some cases this results in a poorer quality material being obtained.

Sublimation

This is the process by which some substances, on heating or cooling, change from solid to gas or vice versa without passing through a liquid state. This property can be used to obtain the substance from dried plant material or a dry crude extract. A simple, small-scale apparatus for this process is shown in Figure 3.6. Plant material is warmed and the vapour allowed to come into contact with a cold surface. The subliming substance vaporizes from the plant material and condenses as a solid on the cold surface, from which it can be scraped off. Caffeine of high purity can

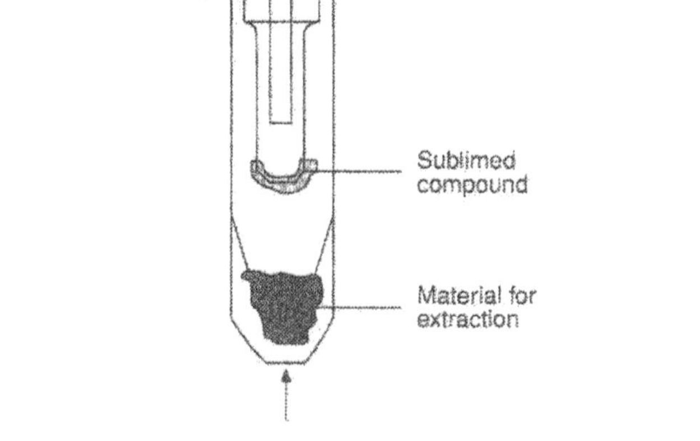

Figure 3.6 'Cold-finger' apparatus for sublimation.

be obtained by this method from dry tea leaves. The process can be refined by using a thermostat to control the temperature to that required for sublimation and by applying negative pressure to encourage movement of the sublimed vapour towards the condensing surface.

Pervaporation

This is a method in which substances dissolved in a liquid phase are removed by binding on to a membrane. Hydrophilic membranes may be used to remove polar materials, including water, from organic solvents and, conversely, hydrophobic membranes can be used to remove organic compounds from an aqueous phase (Karlssen and Tragardh, 1993). This method has been used to remove aroma compounds from fruit juices. This method is currently being developed, and its success will depend on the generation of new membranes which show selective binding for particular chemical groups, perhaps even for specific enantiomers.

Microwave treatment

Enhanced recovery of certain secondary metabolites can be achieved if the plant material is treated with microwave irradiation before or during an extraction procedure (Starmans and Nijhuis, 1996). It is believed that the heating of water in the material causes generation of steam which can result in opening of the matrix and improved porosity.

3.3.6 Preparation of solvent extracts for screening programmes

A number of academic and industrial laboratories are involved in high throughput screening of extracts from living organisms. In many cases, there is no preconception about the chemical nature of the active compounds that are being sought. Thus decisions have to be taken about which solvents to use in order to prepare the extract. Many of the factors described earlier are brought to bear in making this decision (section 3.2). One approach may be to extract the organism exhaustively with a range of solvents of increasing polarity, using a Soxhlet apparatus. Thus four or more extracts will be generated from each sample. If this is felt to be too many for screening (where thousands of samples are available for processing), a simpler approach can be adopted. For instance, only two solvents may be used – one polar, e.g. 70% methanol, and one non-polar, e.g. hexane or light petroleum. If immediate solubility (without organic modifiers) in an aqueous-based bioassay is essential, then this may be further reduced to a polar extract only. At some point, a compromise must be made between what is desirable (i.e. to test every possible type of compound in the organism) and what is practical (i.e. the resources can only support testing of one or two extracts per sample). A decision must also be made on the desirability of applying heat during the process of extraction.

A popular method of extraction for application in a screening programme is to pack a small quantity of the powdered organism to be extracted into a short column with a filter at the lower end, and pass the solvent of choice through this (percolation). The eluant is concentrated if necessary or allowed to collect directly in the wells in which the assay is to take place. A clean-up procedure, e.g. to remove polyphenols (section 3.5.3), may be applied.

3.4 EXTRACTION METHODS FOR SPECIFIC PHYTOCHEMICAL GROUPS

As indicated earlier in this chapter (sections 3.2, 3.3) an important factor governing the general and specific method used in an extraction is the type of chemical class that one is aiming to extract. In this section further details are given on extraction of specific classes of phytochemical. The main groups of compound to be considered are fixed oils, fats and waxes, volatile or essential oils, carotenoids, alkaloids, glycosides, aglycones, phenolic compounds, polysaccharides and proteins. Solvent extraction is the most popular method of extraction and Table 3.2 gives a general outline of the solvents that would be appropriate for extraction of these main classes of compounds.

Table 3.2 Types of phytochemicals extracted by different solvents

Polarity	*Solvent*	*Chemical class extracted*			
Low	Light petroleum	Waxes	Fats	Fixed oils	(Volatile oils)
	Hexane	Waxes	Fats	Fixed oils	(Volatile oils)
	Cyclohexane	Waxes	Fats	Fixed oils	Volatile oils
	Toluene	Alkaloids	Fats	Fixed oils	Volatile oils
	Chloroform	Alkaloids	Aglycones		Volatile oils
Medium	Dichloromethane	Alkaloids	Aglycones		Volatile oils
	Diethylether	Alkaloids	Aglycones		
	Ethylacetate	Alkaloids	Aglycones	Glycosides	
	Acetone	Alkaloids	Aglycones	Glycosides	
	Ethanol			Glycosides	
	Methanol	Sugars	Amino acids	Glycosides	
High	Water	Sugars	Amino acids	Glycosides	
polarity	Aqueous acid	Sugars	Amino acids		Bases
	Aqueous alkali	Sugars	Amino acids		Acids

Polarity and pH are two important factors (section 3.2.3). The methods given in this section are general ones based on the common properties of broad classes of phytochemical. Readers are advised, wherever possible, to consult the literature to obtain procedures appropriate to the specific group, or subgroup, of compounds in which they are interested.

It is advisable when extracting a particular type of chemical to check that the process has been successful. A number of general and semi-specific chemical tests are available for this purpose. These are described in more detail in section 8.2.1.

3.4.1 Fixed oils, fats and waxes

Fixed oils and fats (e.g. soya bean oil, theobroma oil) consist mainly of triglycerides of fatty acids and a small proportion of sterols. Fixed oils and fats differ only in their consistency at room temperature (liquid and solid respectively), which is in turn a function of the degree of unsaturation and chain length of the fatty acids (greater saturation and chain length raise the melting point). Waxes consist of a heterogeneous mixture of fatty acids, long-chain alcohols, esters of the two and paraffins. Fixed oils, fats and waxes are all non-polar in nature and can be extracted efficiently and somewhat selectively using solvents such as light petroleum or hexane. They may also dissolve in chloroform, ethanol or methanol, but these solvents will also extract out other types of phytochemicals. On an industrial scale, these substances may be obtained by the process of expression (section 3.3.5) rather than solvent extraction. Where the presence of fats and oils interferes with the extraction of other constituents, a defatting step involving solvent extraction of fats is sometimes useful (section 3.5.5).

3.4.2 Volatile (essential) oils

Volatile oils are mainly composed of mono- and sesquiterpenes and phenolic compounds. These are also relatively non-polar and may be extracted into light petroleum. However, slightly more polar solvents such as chloroform and dichloromethane are usually more appropriate, particularly where the co-extraction of fats and waxes is to be avoided. Steam distillation is a popular method for the selective extraction of volatile oils, while the Advanced Phytonics method, supercritical fluid extraction and extraction into fixed oils or waxes (see Enfleurage above) are other options (section 3.3).

3.4.3 Carotenoids

Carotenoids are responsible for many of the red, orange and yellow pigments observed in the plant and animal kingdoms. They are generally tetraterpenoid derivatives (containing about 40 carbon atoms) and can be divided into hydrocarbons and oxygenated forms known as xanthophylls. The former group are less polar and can be extracted into petroleum ether. Xanthophylls are more polar as they contain alcohol, ketone, aldehyde, acid or epoxide groups, and can therefore be extracted into ethanol or mixtures of ethanol and less polar solvents such as chloroform.

3.4.4 Alkaloids

All alkaloids contain at least one nitrogen atom and, in the majority of cases, the compound is basic. This means that salt formation can occur in the presence of acid. This fundamental property of alkaloids is used in their extraction and further clean-up. Two methods may be used for alkaloid extraction. One is to basify the plant material using diethylamine or ammonia (see 2 below) and extract with an organic solvent. The alkaline medium ensures that the alkaloids are in their free base or unionized state. Most alkaloidal bases are of medium polarity and can be extracted using chloroform, dichloromethane or diethylether. A general solvent such as ethanol may also be used. The second method is to treat the plant material with aqueous acid. In this situation, the alkaloids form salts which are ionized and therefore soluble in aqueous media. The alkaloid can be recovered in free base form by basifying the aqueous extract (which deprotonates the alkaloid), followed by extraction into a suitable organic solvent.

Some additional points that apply to special cases are:

1. Use of very low pH may result in hydrolysis of ester groups if present in the alkaloid (e.g. aconite alkaloids).
2. Ammonia can react with certain phytochemicals, forming nitrogenous artefacts or false alkaloids.
3. If acidic groups (e.g. phenolic groups) are present in the alkaloidal molecule, then it is possible that at alkaline pH the molecule will be ionized (at the acid function rather than at the nitrogen), thereby reducing drastically its solubility in organic solvents. Morphine (Figure 3.7) is an example of an alkaloid with a phenolic function.
4. Extraction of alkaloids from very fatty materials such as ergot or nux vomica seeds may require prior defatting of the material by extraction with a non-polar solvent, e.g. light petroleum (section 3.5.5).
5. For plant materials containing large amounts of tannins (e.g. cinchona bark) it may be desirable to use calcium hydroxide as the base (instead of diethylamine or ammonia) as this will complex the tannins and allow easier extraction of the alkaloids.
6. Whereas for most alkaloids, the free base is soluble in less polar organic solvents and the salt is soluble in water, there are some exceptions, e.g. lobeline hydrochloride has appreciable solubility in chloroform and the polyhydroxy alkaloids such as castanospermine are

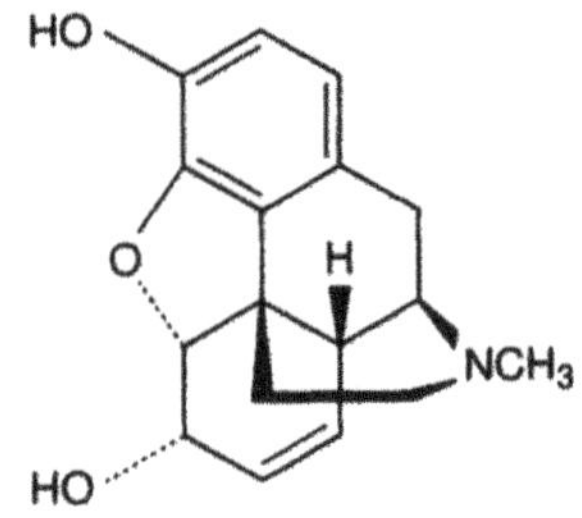

Figure 3.7 Morphine.

water soluble even in free-base form. Thus the standard extraction and solvent–solvent clean-up procedures may not always be applicable.

3.4.5 Glycosides

'Glycoside' is a general term covering a wide range of substances whose common feature is that they consist of at least one sugar molecule linked via its anomeric carbon to another moiety. Technically, the term covers both holosides (e.g. disaccharides, oligosaccharides, polysaccharides), where the second molecule is also a sugar, and heterosides, where the second molecule is a non-sugar. However, in common usage, the term glycoside is taken to refer to heterosides, and the non-sugar moiety referred to as the aglycone or genin. The class of glycoside – anthracene derivative, flavonoid, cardenolide, cyanogenic, etc. – is determined by the aglycone structure (Figure 8.2).

Glycosides are relatively polar due to the presence of one or more sugars in the molecule. The absolute polarity will be determined both by the number and types of sugar (e.g. acetylated or not) and the structure of the aglycone. Most glycosides can be extracted with polar solvents such as acetone, ethanol, methanol, water or mixtures of these. However, cardiac glycosides, with their bulky steroidal aglycones, have appreciable solubility in chloroform. When extracting into water, it is sometimes possible to get enzymic breakdown of the glycoside to the aglycone by the action of glycosidases co-extracted from the plant material. This will not occur if boiling water is used or if significant proportions of alcohol or ammonium sulphate are added to the extract.

In some cases, it may be the aglycone rather than the glycoside that is to be extracted, and this requires hydrolytic separation of the aglycone and sugar before or after extraction. Glycosides in which the aglycone–sugar link is made via an oxygen, nitrogen or sulphur atom (*O*-, *N*-, or *S*-glycosides) can be hydrolysed by boiling with dilute acid (e.g. 10% sulphuric acid). Where there is a C–C link between the aglycone and sugar, the glycoside is resistant to acid hydrolysis, but can be broken down by the action of iron (III) chloride and acid under reflux. However, this is a strong oxidizing agent which may have other effects on the aglycone structure, e.g. oxidation of monoanthrones to anthraquinones.

Aglycones of glycosides may be obtained by hydrolysing extracted glycosides in aqueous media as indicated above, followed by extraction of the aglycone into a less polar solvent, e.g. diethylether or dichloromethane. If the aglycone is appreciably soluble in hot water, then the initial extraction itself can employ warm aqueous acid (which hydrolyses glycosides and extracts aglycones). On cooling the extract, the aglycones become less water soluble and can be partitioned into an organic solvent.

3.4.6 Phenolic compounds

These can exist as free phenols or in glycosidic form. Due to the multiplicity of hydroxyl functions, phenols tend to be relatively polar and dissolve in aqueous alcohols. As they are weak acids, they may also be extracted or partitioned into aqueous alkali as phenolate salts. A problem encountered with phenolic compounds is that they can undergo extensive polymerization reactions by the action of polyphenol oxidases. This reaction is responsible for the development of brown colouration in damaged plant material when exposed to the air and in certain extracts. The polymerization reaction is catalysed by acid.

3.4.7 Proteins

Due to the presence of free carboxylic, amino and phenolic groups on amino acid side-chains in proteins, most can be ionized at high or low pH values. The pH at which no net charge is carried is known as the isoelectric point, pI, and this will vary with each protein depending on the constituent amino acids. At pH values above the pI, the protein carries a net negative charge, and at pH values below the pI, a net positive charge is carried. As a result of this, most proteins can be extracted with water, buffers, dilute acid or base or simple salt solutions. However, more lipophilic proteins require the use of 70–80% alcohols. Table 3.3 summarizes the solubility properties of a range of simple proteins.

Selective precipitation of groups of proteins in a crude protein extract can be achieved by gradual addition of acetone, ethanol or ammonium sulphate. Conversely, for proteins with a greater solubility in salt solutions than water, e.g. globulins, a crude protein mixture can be extracted in 10% sodium chloride solution and the globulins precipitated by addition of water. For prolamines, extraction in 70–80% alcohol can be followed by precipitation by dilution with water. The precipitation step can be followed by resolubilization and further separation using ultrafiltration, gel filtration, ion-exchange chromatography or electrophoresis.

Table 3.3 Solubility properties of simple proteins (adapted from Ikan, 1991)

Protein type	*Solubility characteristics*
Albumins	Water soluble
Globulins	Soluble in dilute salt solutions, water insoluble, precipitate in half-saturating ammonium sulphate
Glutelins	Insoluble in neutral solvents, dissolve in dilute acid or alkali
Prolamines	Dissolve in 70–80% alcohol, insoluble in water, dilute salt solution and absolute alcohol
Scleroproteins	Soluble in strong acids or alkalis, insoluble in water or salt solution
Histones	Soluble in water, dilute acids and alkalis but not ammonia
Protamines	Soluble in water, dilute ammonia, acids and alkali

Proteins vary in their stability to denaturing coagulating agents such as heat, pH, organic solvents, detergents and high pressure. These factors must be taken into account during the extraction process. Where the material to be extracted contains high proportions of fat (e.g. some seeds), it is advisable to defat the material by extraction with light petroleum prior to extraction of proteins.

3.4.8 Polysaccharides

Polysaccharides are polymers of sugars or sugar derivatives. Due to their polyhydroxylated nature, it might be expected that all polysaccharides would be water soluble. However, this is not the case, as the overall shape of the molecule and the strength of intermolecular bonds which have to be overcome for hydration to occur will influence the water solubility.

Generally, there are three types of sugar polymer – those that are completely water soluble, those that partially dissolve in water and swell to form gels and, lastly, those that are water insoluble. Examples of the three types are, respectively, glycogen, amylopectin and cellulose, which are all polymers of glucose. Solubility differences arise as a result of differences in their molecular shape (globular, branched linear and linear, respectively), which are due to differences in the type of linkage between the glucose units. The presence of certain functional groups on the sugar derivatives are also important in determining solubility. For instance, polymers of sugar acids (polyuronides) will dissolve in alkali due to salt formation, whereas the presence of amino sugars would enhance solubility in acid. The acetylamido group in chitin (a major component of the exoskeleton of arthropods) allows strong hydrogen bonding between chitin chains, making this extremely insoluble in water.

Polysaccharides that totally or partially dissolve in water can be extracted using cold or warm water. Co-eluting proteins can often be removed by shaking the aqueous extract with 20% of its volume of chloroform followed by 20% of its volume of *n*-butanol or *n*-pentanol. Proteins will normally coagulate at the interface of the organic and aqueous layers. Acidification of the aqueous layer to pH 4–5 can assist this denaturation process. The crude polysaccharide mixture can be precipitated by the addition of ethanol or acetone to the separated aqueous layer. Divalent ions such as calcium can be used to precipitate out polyuronides, e.g. calcium alginate.

Purification of the polysaccharide fraction can be achieved by redissolving in water and dialysing to remove salts and small molecules, e.g. monosaccharides, followed by freeze-drying to obtain the dried polymer. Individual polysaccharides can be obtained by gel filtration methods.

For water-insoluble polysaccharides such as cellulose and chitin, 'extraction' is achieved by treating material rich in the substance with solvents that will dissolve out the impurities to leave the purified polysaccharide. To obtain cellulose, wood is treated successively with ethanol (to remove resinous material), acidified sodium chlorite (to chlorinate the lignin, bleach), alkali (to elute off chlorinated lignin and remove polyuronide pectins) and finally water (to wash off chemical residues and remaining water-soluble polysaccharides). Each of these treatments may be repeated more than once. The bleaching step may be omitted. Cellulose will dissolve in strong mineral acid solutions from which it can be regenerated by the addition of water. However, all these processes may result in degradation of the cellulose.

3.5 CLEAN-UP TECHNIQUES TO REMOVE UNWANTED SUBSTANCES

3.5.1 Introduction

Many fractionation procedures on extracts shown to have biological activity are aimed at determining the substances responsible, regardless of whether they are novel compounds. However, there are some situations where particular types of compounds must be removed from the extract before further bioassay-linked fractionation takes place.

The materials that may need to be removed fall into two categories:

1. Those present in large quantities, which may render work-up difficult or which may interfere with the test system. The chlorophyll and carotene pigments from leaf material come into this category, as do the brown or red substances often found in bark material from stems and roots. It may be necessary to remove proteins and polysaccharides that are often found in aqueous extracts.

2. Substances that interact non-specifically with a wide range of enzyme systems. The most notorious group in this class are the polyphenolic substances frequently referred to as tannins. These crosslink with many proteins and so inhibit many biological systems. They may, of course, be the only compounds present that are responsible for a reputed or observed biological effect, and in recent years they have attracted considerable interest as bioactive compounds in their own right. However, many research groups that are interested in discovering novel types of molecules will wish to eliminate such compounds at an early stage.

3.5.2 Removal of chlorophyll

Extracts from leaves and other green parts of plants made with ethanol, methanol, chloroform and solvents of similar polarity will contain large amounts of chlorophyll and, when concentrated, form a dark-green viscous paste. Several methods are available for separation of the chlorophyll from other substances present, but care should be taken to check the removed chlorophyll portion for activity in case the process has also removed other components with similar properties to chlorophyll.

Precipitation with lead subacetate – for use with alcoholic extracts

This method relies on the co-precipitaton of chlorophyll with lead ions. It is a classical method for the purification of cardiac glycosides from leaf material, but is not so useful for general fractionation methods as the lead also precipitates alkaloids and phenolic substances such as flavonoids, which may well be the bioactives in a crude extract. An example of the use of this method is given in section 10.2.3.

1. Dilute the alcoholic extract with water so that it is about 20% v/v alcohol : water.
2. Add 10% w/v aqueous lead subacetate solution until no more precipitate forms. (NB If a large amount of chlorophyll and other pigments are present, copious amounts of precipitate are formed and it may be difficult to see if more is formed as more lead subacetate is added. A small amount of the suspension should be centrifuged and the lead subacetate added to the supernatant to check whether more precipitation occurs.)
3. Filter or centrifuge off the precipitate.
4. Add 10% sulphuric acid to precipitate excess lead. Use a sufficient amount so that no further precipitation occurs. NB The acid may cause deterioration of some substances present in the extract so, as an alternative hydrogen sulphide can be bubbled through (CARE!!! TOXIC) to remove the lead.

5. Filter or centrifuge again.
6. Neutralize with 5% barium chloride solution and filter or centrifuge. The filtrate should be colourless or only faintly coloured and can be used for further fractionation.

Solvent partition – for alcoholic extracts

Chlorophyll, although extracted to some extent by alcohol, is more soluble in less polar solvents and so can be removed by solvent partitioning.

1. Dilute the alcoholic extract to about 20% v/v alcohol with water.
2. Extract with several aliquots of diethyl ether (CARE!!! INFLAMMABLE) or, preferably dichloromethane (CARE!!! emulsions may form). The chlorophyll should go into the less polar layer.

NB Other less polar constituents will also partition into the dichloromethane or ether layer so a further step to fractionate this layer using exclusion gel is advisable.

Use of exclusion gel

This technique separates chlorophyll as it is a large molecule and passes through the gel in the column quickly. The other components in the extract are retained. Sephadex LH-20® is used as this is compatible with the non-polar nature of the solvent.

1. Allow the Sephadex LH-20® to swell for some hours in the solvent in which the extract is dissolved (use 10 g Sephadex LH-20® per 100 mg extract).
2. Pack the gel into a column.
3. Introduce the extract, dissolved in as small a volume as possible, to the top of the column.
4. Elute with the same solvent and collect aliquots of about 5 ml. A green band corresponding to the chlorophyll should be seen which moves quickly through the gel. Other yellow bands (corresponding to carotenes) may also be observed.
5. Continue elution until all the pigments have been removed and collect at least twice the column volume of subsequent eluent – this will contain other constituents. Monitor fractions by TLC or bioactivity if necessary.
6. Before further fractionation bulk the fractions giving the same response.

Reverse-phase column chromatography – for alcoholic extracts

This method exploits the non-polar nature of chlorophyll so that it is retained on a reverse-phase adsorbent while more polar components pass through. Because of the expense of reverse-phase materials it is most suitable for small-scale work. However, the material can also be reused after thorough washing with a series of solvents of decreasing polarity.

1. Prepare the stationary phase (RP-8 or RP-18 silica; 25 g/100 mg extract) by treating with methanol, and pack into a column. (NOTE: washed and purified charcoal can be used but this may retain other constituents of the mixture and it may be difficult to remove them.)
2. Wash the column with the solvent in which the sample is dissolved.
3. Dissolve the extract in a small amount of water or methanol : water 50 : 50 and introduce to the top of the column.
4. Elute with 50 : 50 methanol : water followed by methanol (collect fractions if desired). Use at least 100 ml solvent per 100 mg extract. These fractions will contain most of the components of the extract other than the chlorophyll but it is wise to check by using TLC.
5. Elute the column with 100 ml acetone : methanol 1 : 1. This may elute some chlorophyll.
6. Elute with dichloromethane or ether to obtain the chlorophyll.

3.5.3 Removal of polyphenolics

Polyphenolic substances are often coloured brown or red and many oxidize quite rapidly when exposed to air to form insoluble precipitates, particularly in an acidic environment. It is difficult to ensure that all such compounds are removed from an extract and several methods are reported in the literature. A useful paper by Wall *et al.* (1996) summarizes the relative effectiveness of these and concluded that polyphenols were best removed from aqueous extracts by the use of Sephadex LH-20® and from alcoholic extracts by solvent partitioning. Treatment with polyvinylpolypyrrolidone (PVPP), although effective, is not useful for large-scale work. An additional problem is the binding of a number of non-phenolic substances to PVPP, resulting in their removal from the extract.

The different techniques can be summarized as follows:

Preferred technique for aqueous extracts: Sephadex LH-20®

Use 10 g Sephadex LH-20® per 100 mg extract.

This method relies on the fact that polyphenolic compounds are retained by the Sephadex partly on account of their size but also because of hydrogen bonding with the gel molecules.

1. Soak the gel in ethanol for several hours and then pack into a column.
2. Pass a sufficient volume of ethanol : water 4 : 1 through the column to replace the ethanol.
3. Apply the extract dissolved in ethanol : water 4 : 1 (100 mg extract per 1.5 ml solvent) to the column.

4. Elute column with ethanol (100 ml per 100 mg extract).
5. Elute with acetone : water 1 : 1 (50 ml per 100 mg extract).

The ethanol eluate contains the extract free from polyphenols, the acetone : water extract contains polyphenolics.

Preferred technique for alcoholic extracts: solvent partition

In this method very non-polar substances are removed with hexane and the polyphenols by water so that other compounds remain in the chloroformic layer which is washed with 1% saline to remove any traces of polyphenols.

1. Add water to an extract made with methanol or ethanol to make 90% v/v alcohol extract.
2. Partition with an equal volume of hexane.
3. Discard the hexane layer unless the non-polar components are required for further study.
4. Concentrate the remaining methanolic layer under reduced pressure and partition the residue with equal volumes of chloroform : methanol 4 : 1 and water.
5. Separate the two layers and wash the chloroformic layer with an equal volume of 1% w/v NaCl in water.

The chloroformic layer will contain the polyphenol-free extract.

NB Ethyl acetate has sometimes been used instead of chloroform : methanol 4 : 1 but this is NOT ADVISABLE since some polyphenols are extracted into the ethyl acetate layer.

Use of polyamide – useful for small-scale work

Use 5 g polyamide per 100 mg of extract.

This technique relies on the complexation of polyphenolics with polyamide.

1. Soak the polyamide (column chromatography grade) in water overnight and pack into a column.
2. Dissolve the extract in a minimum volume of methanol and apply to the column.
3. Elute column with methanol (about 200 ml per 5 g polyamide) or until the eluate is clear.

The polyphenolics are retained on the column, the eluate contains the extract free from polyphenols.

Polyvinylpolypyrrolidone (PVPP)

Use 2.5 g PVPP per 100 mg extract.

This technique relies on the same principles as the use of polyamide (above).

1. Dissolve extract in water (30 ml per 100 mg extract) – or methanol : water 1 : 1(30 ml per 100 mg extract) for less polar extracts.
2. Add the PVPP and stir well. Allow to stand for 5 min with regular stirring.
3. Centrifuge (or filter).

The supernatant (or filtrate) should be free of polyphenolics.

NB Purified collagen or hide powder can be used instead of PVPP.

Use of Whatman DE52 resin

Use 10 g resin per 1 g extract.

1. Dissolve extract in water.
2. Add the DE52 resin and stir well. Allow to stand for 5 min with regular stirring.
3. Centrifuge or filter.

3.5.4 Removal of proteins

Proteins are commonly removed from aqueous extracts by the addition of aqueous ammonium sulphate to give a solution of increasing strength, i.e. 5%, 10%, 20%, etc. At each stage particular proteins precipitate out and they can be filtered off. Some types of proteins can be removed by precipitation by changing the pH of the solution or adding salts. Seed storage proteins can be removed from an extract made with water by lowering the solution to pH 5 by the addition of acetic acid. Large protein molecules can also be removed by dialysis (section 4.3.4).

3.5.5 Removal of fats and oils

Fats and oils are often not very active biologically but their presence in material may reduce the extraction of other constituents by more polar solvents and may interfere with bioassays. The removal of fats and oils is a common procedure when seeds and fruits are extracted, but may also be necessary with other materials, such as animal matter and waxy leaves from plants. Several methods can be used.

Solvent extraction – 'defatting'

This technique depends on the high solubility of fats in very non-polar solvents compared with most other plant constituents.

1. Exhaustively extract the powdered or chopped material with light petroleum or, less preferably, dichloromethane.
2. Filter off the plant material for further extraction and studies.

The light petroleum extract will contain fats and waxes. (NOTE: the light petroleum extract may contain other non-polar constituents, particularly if volatile oils are present in the material, so should always be checked by chromatography and by the relevant bioassay to ensure that only non-active substances are present.)

Partition into a fat which solidifies

This is a variation of the previous procedure and is useful for aqueous or dilute alcoholic extracts. An inert fat is added which melts at a temperature below the boiling point of the extraction fluid.

1. Add hard paraffin wax (pharmaceutical quality) or carnauba wax (pharmaceutical quality) to the extract (or the plant material in the extracting solvent) contained in a wide-mouth beaker. Use 10 g wax for every 25 g plant material.
2. Heat the mixture until all the wax melts and stir the mixture continuously.
3. Remove the heat and allow to cool to room temperature. The wax will solidify as a hard layer on the top of the extract.
4. Make two holes in the hard wax layer and carefully pour out the liquid underneath through one of the holes, using the other for air to enter.

This liquid should be free of fats which have all dissolved in the wax.

3.5.6 Removal of water-soluble polysaccharides and other carbohydrate polymers

Many biological samples contain appreciable amounts of sugar polymers, such as starch, mucilage, gums, etc., which can hinder extraction and fractionation since they form viscous solutions or gels with water. They occur in high amounts in marine algae, in the food storage organs of the more highly evolved plants and in certain animals.

The most common method of removing these substances from aqueous extracts is to precipitate them by the addition of a less polar water-soluble solvent, such as ethanol or acetone, to the water. The resulting precipitates can then be removed by filtration or centrifugation.

3.5.7 Removal of sugars

High concentrations of sugars in a water extract can cause problems when the extract is concentrated because of the viscous syrups formed and the possibility that heating or acid treatment may result in charring.

Reduction in solvent polarity

Sugars are not very soluble in solvents with a lower polarity than water, so can sometimes be precipitated from aqueous extracts by addition of methanol, ethanol or acetone. The precipitates formed are not always very easy to remove by filtration or centrifugation.

Exclusion chromatography

Sugars are quite small molecules so can be separated using standard hydrophilic exclusion gels such as Sephadex® (10 g Sephadex® per 100 mg extract). The sugars are retained by the column and elute last, while the larger molecules pass through and are eluted first. Sephadex G10® is suitable for this purpose. The eluant should be collected in fractions which can be monitored by TLC to identify the sugar and de-sugarized components.

Reverse-phase chromatography

Sugars are usually very polar so are not retained on a reverse-phase (RP) silica stationary phase when water (or especially aqueous alcohol) is used as an eluant. Use 25 g stationary phase per 100 mg extract. The other, less polar components of the mixture can be eluted from the column by decreasing the polarity of the eluant, e.g. by use of methanol or methanol : acetone mixtures. (NOTE: washed and purified charcoal can be used but this may retain other constituents of the mixture and it may be difficult to remove them.)

3.6 BIBLIOGRAPHY

Ikan, R. (1991) *Natural Products – A Laboratory Guide*, 2nd edn, Academic Press, London.

Karlsson, H.O.E. and Tragardh, G. (1993) Pervaporation of dilute organic water mixtures. A literature review on modelling studies and applications to aroma compound recovery. *J. Membr. Sci.*, **76**, 121–46.

King, M.B. and Bott, T.R. (eds) (1993) *Extraction of Natural Products using Near-critical Solvents*, Blackie, Glasgow.

Modey, W.K., Mulholland, D.A. and Raynor, M.W. (1996) Analytical supercritical fluid extraction of natural products. *Phytochem. Anal.*, **7**, 1–15.

Starmans, D.A.J. and Nijhuis, H.H. (1996) Extraction of secondary metabolites from plant material: a review. *Trends Food Sci. Technol.*, **7**, 191–7.

Wall, M.E., Wani, M.C., Brown, D.M. *et al.* (1996) Effect of tannins on screening of plant extracts for their enzyme inhibitory activity and techniques for their removal. *Phytomed.*, **3**, (4), 281–5.

Wilde, P.F. (1994) New solvents for extraction. *Perfume, Flavour*, **19**, 25–7.

Williamson, E.M., Okpako, D.T. and Evans, F.J. (1996) *Selection, Preparation and Pharmacological Evaluation of Plant Material*, Wiley, Chichester.

ADDRESS

Advanced Phytonics, Olway Works, Healey Road, Ossett, WF5 8LT, UK
Tel.: +44 (0)1924 261 281; +44 (0)1924 264 645.

CHAPTER FOUR

Crude fractionation procedures

4.1 INTRODUCTION

The components of a mixture, such as an extract from a living organism, can be separated into groups of compounds sharing similar physicochemical characteristics. This process is called fractionation and can be carried out in various ways, each of which groups compounds according to one or more particular features. Thus solubility, size, shape, electrical charge and several other features may influence the grouping. Note that the molecules grouped together in a fraction separated according to one method may not be the same as those combined in a fraction obtained by a method based on a different criterion.

The skilful application of fractionation methods exploits these differences so that, when two methods are used in sequence, many fractions can be obtained, each containing only one or two components. Thus the initial fractionation may be based on solubility differences while the second may utilize molecular size.

Although in many situations all components of a mixture will need to be tested, in some situations the extract will contain large amounts of unwanted compounds which should be removed, e.g. chlorophyll from leaf extracts, oils from seed extracts, sugar material. The removal processes to yield an extract where the constituents of interest remain are called clean-up procedures. Clean-up procedures use many of the same techniques used in fractionation, and methods for removal of commonly encountered contaminants are described in section 3.5.

Methods used for achieving fractionation or clean-up are listed below and are all of use:

- precipitation;
- solvent–solvent extraction;
- distillation;
- dialysis;
- chromatographic procedures; and
- electrophoresis.

It should be noted that chromatography in its various forms is the most common method and is exploited in most fractionation procedures carried out nowadays. The principles underlying chromatography and the various techniques employed are therefore considered in some depth in Chapters 5, 6 and 7.

4.2 CHOICE OF FRACTIONATION METHOD

The choice of fractionation method to be used in any particular situation depends on several factors.

4.2.1 The nature of the substances present in the extract

This is probably the most important point to be considered. Unless preliminary chemical tests have been carried out (section 8.2.1), this knowledge may consist of only the approximate solubility of the components based on the type of solvent used to make the extract. Thus if water is used as the extracting solvent, the components present will be polar in nature and may include compounds carrying an electrical charge. On the other hand, if a non-polar solvent such as hexane has been used, mainly non-polar, unionized substances will form the major components of the mixture. Another factor concerned with the components of the mixture is their susceptibility to degradation while going through the separation process. The stability of substances present is often not known, so a good general principle is to carry out fractionations under the mildest possible conditions, i.e. minimization of temperature, protection from light and avoidance of reactive solvents and other chemical substances.

4.2.2 The immediate fate of the separated fractions

If the fractions are to be used for biological tests, then toxic materials should be avoided for the fractionation process and the fractions should be dissolved in solvents compatible with the test system. This usually means an aqueous medium of some sort. The aspect of toxicity is less important if the fractions are to be used for further fractionation procedures (where the toxic materials may be removed) or for isolation of individual chemical substances.

If the fractions are to be concentrated prior to the next stage of the investigation, then it may be useful to ensure that the concentration stage is facilitated by using a volatile solvent if at all possible.

4.2.3 Availability and cost of equipment and materials required

Some sophisticated chromatographic media and equipment can be very expensive, but much cheaper alternatives can frequently be used. With a little skill, efficient fractionation can be carried out using quite simple procedures and material. Such alternatives will be highlighted in this book to assist those with constraints due to lack of availability of materials or finance.

4.2.4 Safety

Most chemical manipulations incur some degree of risk. This may be the acute or chronic adverse effects on the health of personnel because of toxicity, or possible material damage caused by such factors as fire or corrosion. Techniques and materials should always be chosen which reduce these risks to a minimum. Thus non-flammable solvents are always used in preference to flammable ones. Many databases now exist which provide information on the hazards of chemical substances, and these should always be consulted before a new procedure is carried out.

4.3 COMMON FRACTIONATION METHODS

4.3.1 Precipitation

Precipitation occurs when the concentration of a substance in a solution exceeds its maximum solubility. A compound can be precipitated by a variety of methods, each dependent on altering one of the factors influencing solubility. Precipitation can be used either to remove the substances of interest or to remove unwanted material and retain that of interest in solution.

A simple method of achieving precipitation is to lower the temperature of the solution of the extract. The less soluble components will precipitate out and can be separated by centrifugation or filtration. If one substance is present as a major constituent, it may precipitate out as crystals. This technique is widely used as a final stage in the isolation of single compounds from aliquots derived from the last fractionation stage.

The usefulness of precipitation by cooling is somewhat limited in the case of aqueous solutions. If the extract is originally at room temperature, only a small reduction in temperature is possible before the whole solution freezes.

Another way to effect precipitation is to alter the polarity of the solvent by the addition of a miscible solvent of different polarity. If the original extract has been made with water, then methanol, 96% ethanol or acetone can be added in aliquots so that, with each addition, the solvent becomes less polar and the more polar solutes precipitate. Conversely, if a non-polar solvent such as hexane is used for the initial extraction, the addition in the same way of a miscible, comparatively polar solvent such as acetone will precipitate out the least polar components.

Another approach with aqueous extracts is to add a strong solution of a very water-soluble electrolyte. The less-soluble non-ionic compounds present in an extract will be displaced and will be precipitated. This technique is known as 'salting out' and is widely used in the preparation of the protein and peptide components from extracts. Ammonium sulphate is one of the compounds most commonly used in this context.

A common reason for precipitation, often encountered in classical chemistry, is the complexation between a compound in the extract and one in the solution when a reagent is added. This method has some applications in the separation of plant constituents but it should be noted that any complex formed should be easily reversible to yield the compound originally present in the plant extract.

Some reagents used to precipitate different types of natural products are shown in Table 4.1.

It should be noted that, when precipitates are formed they are often composed of very fine particles and this causes problems in filtration since the filter pores easily become blocked. Filtration can be improved by using a filter bed of washed filtration-grade kieselguhr made with negative pressure, mixing more kieselguhr with the extract and precipitate and then pouring on to the prepared filter bed. The bed is washed with the solvent used to prepare the extract to ensure that no trace of the extract remains. The precipitate can then be dissolved out of the filter bed matrix with an appropriate solvent (Table 4.1). This technique is described in section 9.1.1. Alternatively the precipitate can be removed by centrifugation.

4.3.2 Solvent–solvent extraction

Immiscible liquids

When a liquid is added to an extract dissolved in another liquid which does not mix with the first, two layers are formed. Any component of the mixture will have a solubility in each of the two layers (usually called phases) and after a time an equilibrium of concentrations in the two layers is reached. The time taken for equilibrium to be achieved is usually shortened by fairly vigorous mixing of the two phases in a stoppered separating flask (Figure 4.1).

CAUTION!! It is extremely important that volatile solvents are not mixed with hot (or even warm) aqueous phases. This will cause a large increase in vapour pressure resulting in the stopper being blown off and the contents of the separating funnel being sprayed out. This may also

Table 4.1 Precipitation reagents

Precipitation complexation reagent	*Phytochemical type precipitated*	*Conditions for reversing complex*
Lead subacetate solution 10%	Flavonoids, chlorophyll, polyphenolic compounds	Treat with aqueous alkali
Calcium hydroxide (in paste form)	Polyphenolic compounds (tannins), alkaloids	Treat with excess dilute acid Remove alkaloids from paste with comparatively non-polar solvent
Picric acid	Alkaloids	Treat ppt with aqueous alkali and extract alkaloid with dichloromethane or similar solvent
Dragendorff's reagent (potassium iodobismuthate)	Alkaloids	Dissolve ppt in acetone : methanol : water 6 : 2 : 1 and pass through anion-exchange column
Mayer's reagent (potassium iodomercurate)	Alkaloids	Dissolve ppt in acetone : methanol : water 6 : 2 : 1 and pass through an anion-exchange column
Ammonium reineckate	Alkaloids	Dissolve ppt in acetone : methanol : water 6 : 2 : 1 and pass through an anion-exchange column
Powdered hide	Polyphenolic compounds (tannins)	Treatment with aqueous alkali
5% Gelatin solution	Polyphenolic compounds (tannins)	Treatment with aqueous alkali
Polyvinylpolypyrrolidone (PVPP)	Polyphenolic compounds (tannins)	Treatment with aqueous alkali: elute with methanol

ppt, Precipitate.

occur with cold fluids if an exothermic reaction takes place, e.g. mixing of acids and bases, dilution of strong acids.

CAUTION!! Some organic phases easily form emulsions with aqueous phases, particularly if small particles are present or are formed by precipitation. The chlorinated solvents chloroform and dichloromethane are very likely to produce these emulsions. The formation of emulsions is best avoided by not using these solvents or, if they are used, mixing the two phases very gently. If emulsions are produced, they often form an

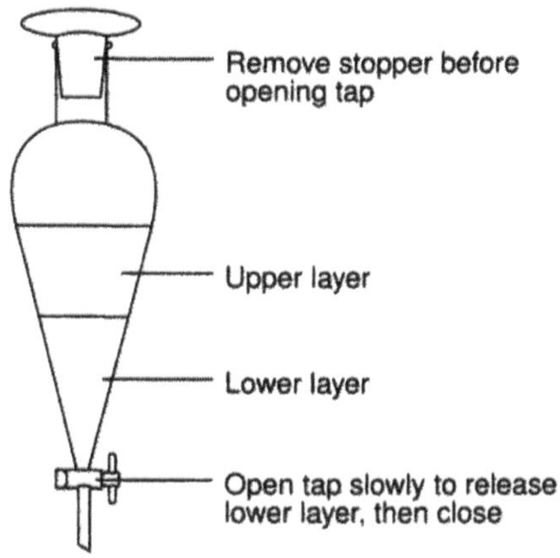

Figure 4.1 Separating funnel.

opaque third layer at the interface of the two clear phases. Some methods of resolving emulsions ('cracking') are described below.

The solubility of an uncharged substance in any phase at a fixed temperature is dependent on its similarity in polarity to the liquid phase, the principle 'like dissolves like' applies. Charged molecules have a high affinity for liquids where a large number of ions with opposite charge exist and so, in this case, 'opposites attract', e.g. acidic substances will dissolve more in a basic than a neutral or acidic aqueous phase. The ratio of concentrations of a substance in the two phases is called the partition coefficient, κ. Different substances will have different partition coefficients so that, if one compound is fairly polar, its partition coefficient relative to the more polar phase is much higher than that of a non-polar compound.

Thus, when an extract is exposed to two immiscible liquids, the different solutes will distribute themselves differently. If the partition coefficient is greater than 100 for any substance, a large proportion of it will be present in only one of the phases.

Cracking emulsions

Run the emulsion into a clean separating funnel, thus separating it from the other phases. The following treatments can then be tried:

1. Gently twist the separating funnel in the vertical plane and tap gently on the sides with fingers or a blunt, non-fragile rod.
2. Addition of a small amount of methanol or ethanol followed by treatment as in (1). The alcohol reduces the interfacial tension between the two phases and thus enables them to separate more easily.
3. Filtration through commercially available silicone-treated filter paper

under reduced pressure. This allows passage of the non-polar phase but not the water component.

4. If small particles are present, filtration under reduced pressure can be tried. The particles are often so small that they would quickly block the pores in ordinary filter paper, so the formation and use of a kieselguhr bed for filtration is usually preferred (section 9.1).

Multiple extraction

Further fractionation of a substance into only one phase can be achieved by successive extraction of the original phase in question with the opposite phase. It is better to use several successive elutions of relatively small volume than one amount of the total volume.

This is exemplified by the following: an aqueous extract (100 ml) contains a compound, A, with partition coefficient κ (chloroform : water) 4 : 1. The extract can be extracted either with four successive 25 ml aliquots of chloroform or by one 100 ml aliquot.

$$\text{Partition coefficient } \kappa = \frac{[A]_{CHLOROFORM}}{[A]_{WATER}} = 4$$

Extraction with 4 × 25 ml aliquots

After equilibrium $[A]_{CHLOROFORM} = x/25$

$$[A]_{WATER} = (100-x)/100$$

From κ above

$$x/25 = 4[(100-x)/100]$$

$$100x = 4 \times 25(100-x)$$

$$100x = 10\,000-100x$$

$$200x = 10\,000$$

$$x = 50$$

i.e. 50% of A is removed from water phase into the chloroform. So, after four extractions, 50% of original + 50% of 50% (25%) + 50% of 25% (12.5%) + 50% of 12.5% (6.25%) is removed = **93.75%** removed after four extractions.

Extraction with one 100 ml aliquot

[A] in chloroform = 4[A] in water (since both are 100 ml) = **80.0%** removed into chloroform, i.e. the extraction is not so efficient as using 4 × 25 ml aliquots of chloroform.

Successive liquid extractions can be carried out with immiscible liquids with a range of polarities. Thus a water extract can be extracted successively with hexane, dichloromethane, ethyl acetate (or dichloromethane : ethanol

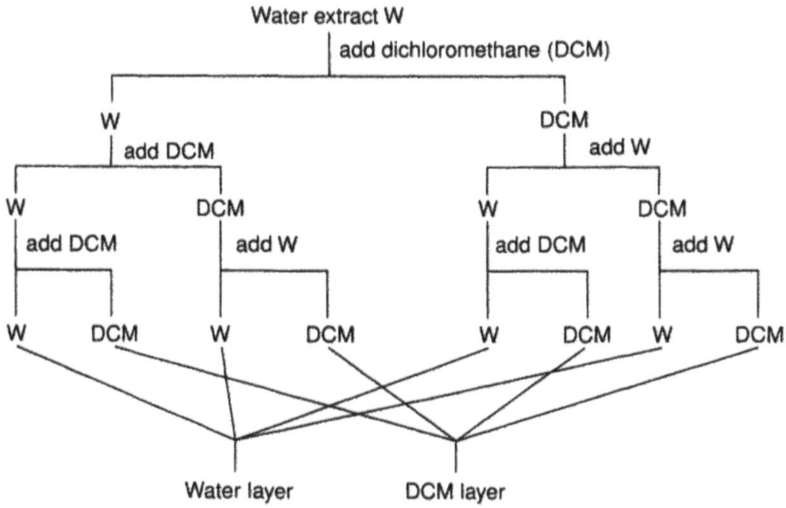

Figure 4.2 Scheme for multiple liquid–liquid fractionation.

3 : 1) and butanol. The extracting solvents of this series are of gradually increasing polarity. Conversely, a petrol or dichloromethane extract may be extracted with a solvent series of decreasing polarity, e.g. water, water : ethanol 3 : 1, water : ethanol 2 : 1, water : ethanol 1 : 1. It is usual to extract each phase several times with the other one and to combine the like phases. A typical scheme is shown in Figure 4.2.

Salting out

The extraction of moderately polar compounds from aqueous extracts by somewhat immiscible solvents such as butanol or methyl ethyl ketone can be enhanced by 'salting out'. The solubility of many non-ionic compounds in water is considerably reduced when a large amount of a very soluble inorganic salt is added to the solution. The non-ionic compound may precipitate out if no other solvent is present but, in the presence of another moderately polar solvent, it will usually dissolve into that one. Common salt, sodium chloride, is often used for this purpose because it is cheap and freely available, but compounds such as ammonium sulphate are also widely used. The addition of 1 g/ml sodium chloride to water will render polar solvents such as methanol and acetonitrile immiscible and allow them to be used for solvent–solvent extraction.

Partitions using changes in pH

Acidic and basic fractions can be obtained and purified from extracts by exploiting the pH-dependent changes in water solubility of relatively non-polar acidic and basic substances. As noted above, the ionic forms of these are soluble in an aqueous media whereas the non-ionized form is soluble in less polar solvents. Thus naturally occurring nitrogenous bases (the majority of which are known as 'alkaloids') can be extracted from a non-polar extract made with dichloromethane by dilute acid, since the nitrogen in bases becomes protonated and the molecule carries an electrical charge and is much more water soluble. However, the acidic aqueous layer will also contain non-basic water-soluble compounds. The basic components can be separated from these by a further step consisting of basification of the acidic layer and extraction with a fresh amount of dichloromethane. The basification process converts the protonated bases back to the non-polar uncharged form which is dissolved preferentially in the dichloromethane and leaves the non-basic polar components of the mixture in the aqueous phase (section 3.4.4). This approach is very commonly used in the production of crude alkaloidal extracts from plant material and is outlined in Figure 4.3.

A similar process is used to separate acidic and phenolic substances as a fraction from a general aqueous extract. Acidification of the water layer in this case causes the compounds in question to dissolve preferentially in the organic phase whence they can be separated from non-acidic, non-polar material by extraction with aqueous base.

Removing traces of water from organic solvents

It should be realized that the extraction solvent will also extract small amounts of the other solvent. When aqueous and non-polar, sometimes called 'organic', solvents form the two phases, the resulting organic layer

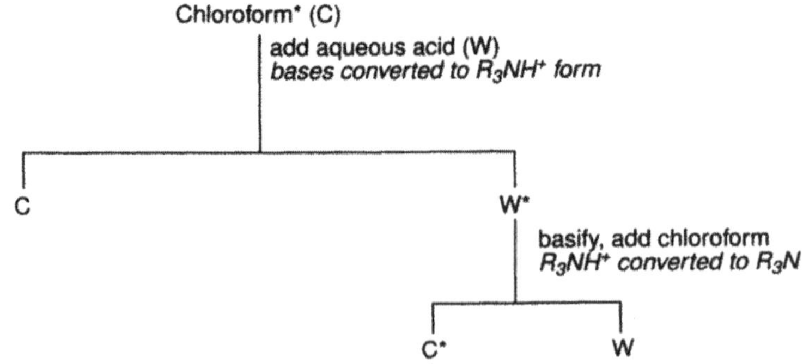

Figure 4.3 Extraction of a fraction containing non-polar ionizable basic compounds. *, Fraction containing basic substances.

often contains small amounts of water after separation from the aqueous phase. The organic layer usually has a faintly cloudy appearance and can be dried by the addition of anhydrous sodium sulphate. When the salt is added, with good mixing by agitation or swirling, it can be seen to appear more crystalline and the organic solvent often appears much more transparent. Each molecule of anhydrous sodium sulphate takes up 10 molecules of water of crystallization and so this process absorbs the water and leaves the organic solvent 'dry'. The sodium sulphate should be added in successive moderate amounts until no more crystalline material is seen to form. The solid residue can then be filtered off and traces of water removed by the use of filter paper treated with silicone (which does not permit passage of water) or by filtration through absorbent cotton (cotton wool) where the cellulose of the cotton fibres absorbs large amounts of water.

4.3.3 Distillation

The separation of a mixture of volatile compounds can be effected by fractional distillation. This process is used extensively in the petrochemical industry but has very limited applications in the fractionation of plant extracts and can only be used for volatile (sometimes called 'essential') oils.

Fractional distillation relies on the temperature gradient established along a cooling column situated above a boiling liquid. The least volatile compounds condense nearest the boiling chamber whereas the most volatile condense much further away. Withdrawal points are positioned along the column and the liquid condensate taken from each will differ in composition (Figure 4.4).

4.3.4 Dialysis

Dialysis is a method of separating the components of a mixture according to their molecular size. This process occurs naturally across cell membranes and is very important in many physiological processes.

The essential part of any dialysis procedure is a thin, semipermeable membrane, consisting of a polymeric material with regular pores, which allows the passage of small molecules (molecular mass < 1000 daltons). The passage of larger molecules is considerably impeded or not possible. Under pressure or an osmotic gradient nearly all the small molecules in a mixture pass through the membrane, leaving behind the larger molecules.

Products having a range of pore sizes are available as commercial dialysis membranes so that, by the use of successive membranes with increasing pore size, a mixture may be fractionated according to the size ranges of the molecules present.

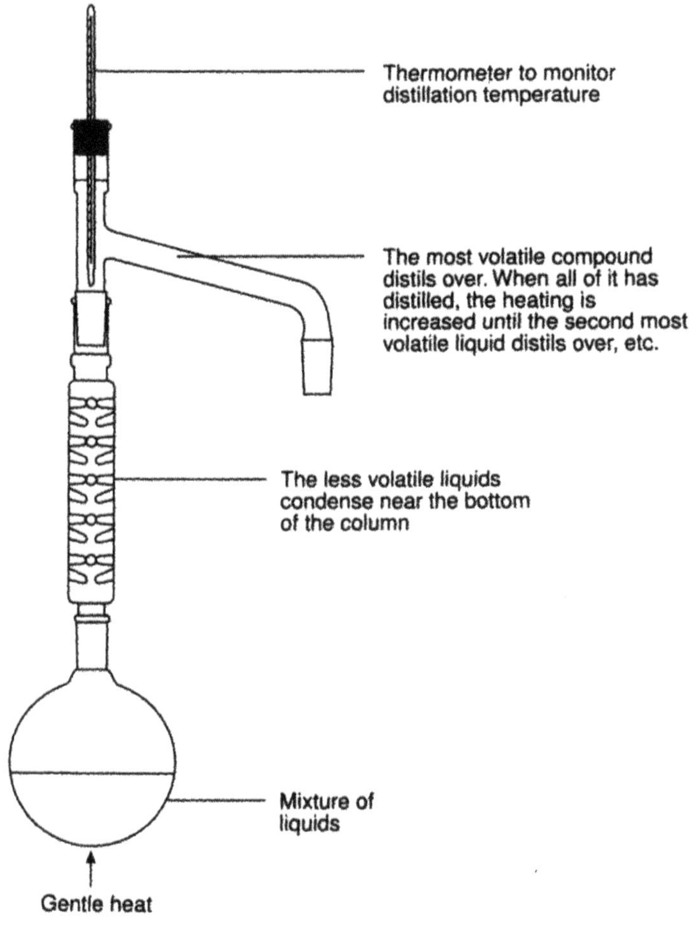

Figure 4.4 Fractional distillation column.

Dialysis is often used in the purification of polysaccharides and proteins in a process sometimes called 'desalting'. The aqueous extract containing the large molecules is enclosed in a membrane and immersed in water. Low molecular mass compounds such as salts, sugars and other substances pass out through the membrane, while water molecules pass the other way through the membrane because of the osmotic gradient. The large molecules retained within the membrane are therefore purified of small molecular mass components but the solution becomes more diluted. Conversely, the same process can be used to remove large molecular mass substances if small molecules are the focus of interest. To achieve complete removal of small molecules the aqueous phase outside the membrane must be replaced periodically, since the movement across the membrane will only take place until the concentration on both sides is equal.

4.3.5 Electrophoresis

Electrophoresis is a method of separating a mixture of substances that carry an electrical charge. Under the influence of an electrical field the constituent molecular species will move at different rates according to their size, shape and total electrical charge. Electrophoresis is used primarily as an analytical method for small samples of mixtures of charged molecules, particularly proteins, peptides and amino acids, rather than as a fractionation procedure. It can be used in a similar way to preparative layer chromatography (section 7.2) to separate the components of fairly simple mixtures.

There has been a considerable revival of interest in analytical electrophoresis in recent years due to the introduction of the technique known as capillary zone electrophoresis (CZE). This technique has the advantage over earlier ones since it can be modified to analyse uncharged molecules as well as charged ones. This technique has not yet been adapted for preparative applications.

4.4 BIBLIOGRAPHY

Harborne, J.B. (1984) *Phytochemical Methods*, 2nd edn, Chapman & Hall, London.
Ikan, R. (1991) *Natural Products – A Laboratory Guide*, 2nd edn, Academic Press, London.

CHAPTER FIVE

Chromatographic fractionation procedures

5.1 INTRODUCTION

Chromatographic procedures are the most diverse and the most widely used techniques in the fractionation of extracts. There is no doubt that they have made possible the isolation of many naturally occurring compounds.

The chromatographic techniques used in fractionation are those used in preparative chromatography. It should be pointed out that the major use of chromatography is as an analytical method for the quantitation of one or more compounds in a mixture without the need for very much of a clean-up procedure (section 8.3.4). The techniques discussed here are based on the same principles but are scaled up and adapted for the isolation of quantities of at least 10 mg of a substance.

All chromatography relies on the differential distribution of compounds between two phases, one of which moves relative to the other. These phases are called the mobile and stationary phase, respectively. The mobile phase is a fluid which can be either a liquid, a gas or a supercritical fluid, but in practically all the fractionation procedures described below it is a liquid. The stationary phase as used usually appears to be a solid consisting of fine particles. In some cases the solid material is in fact the phase that takes part in the chromatographic process, but in other cases the stationary phase, as far as this process is concerned, is a monomolecular film of 'liquid' bonded to the solid support material or a thicker film of liquid coated on it.

5.2 THE PRINCIPLES OF CHROMATOGRAPHY

5.2.1 The chromatographic process

Chromatography is based on the fact that a dynamic equilibrium is established between the concentration of a solute in two phases. This dynamic equilibrium consists of molecules of the solute continually passing between the two phases. At any one time the concentration ratio, i.e. ratio of number of molecules in each phase, is constant. However, each molecule spends time in both phases and the proportion of time spent in a phase depends on the relative attractions of the substance for the two phases. The concentration ratio at equilibrium is called the distribution coefficient and the name for the attraction for any phase is called the affinity, i.e. a compound that moves more slowly is said to have a higher affinity for the stationary phase than one that moves more quickly.

The chromatographic process is conceptualized as a succession of equilibria established in a series of very small units of both phases along the path that the mobile phase travels (Figure 5.1). The name given in chromatographic theory to this small unit is the theoretical plate, a rather misleading term which has arisen from the concept of fractional distillation.

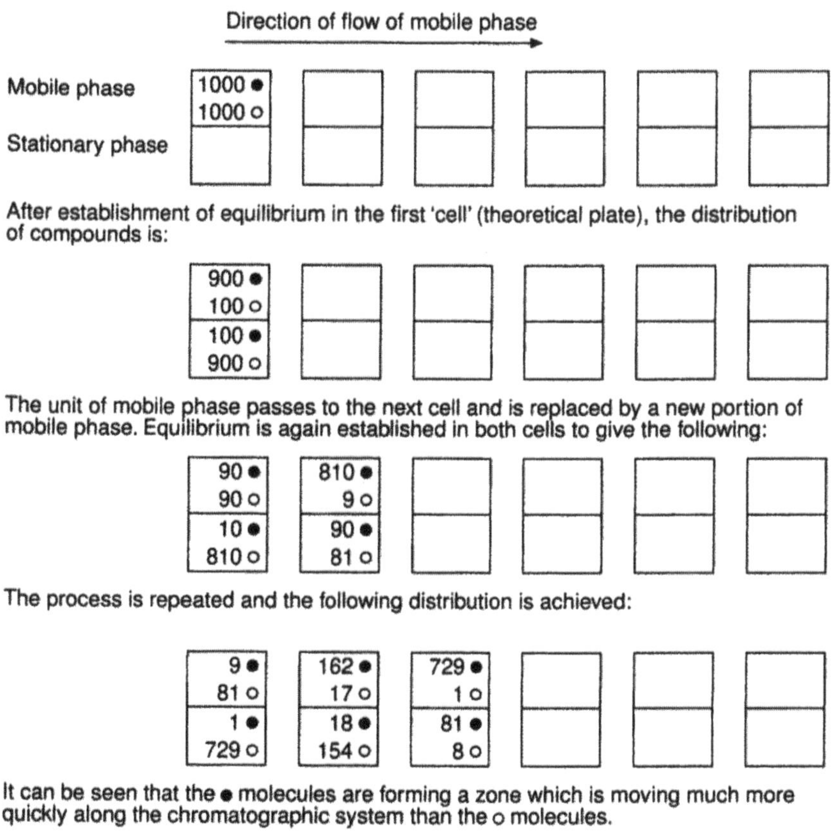

Figure 5.1 Chromatographic separation of two compounds. A mixture of two types of molecule, ● and ○, dissolved in the mobile phase is introduced to the system. The way in which 1000 molecules of each type distribute themselves over a period of time is illustrated. The distribution coefficients (κ = [mobile phase]:[stationary phase]) of the two substances differs. For example, κ ● = 9 : 1; κ ○ = 1 : 9, i.e. ● has a greater affinity for the mobile phase than ○.

Table 5.1 Types of intermolecular force involved in chromatographic processes

Intermolecular force	*Typical groups in molecule*	*Relative strength of intermolecular force*
Van der Waals interactions	Hydrocarbon chains	Weak
Dipolar and induced dipolar attractions	C=C double bonds, Halogens, nitrogen	Weak
Hydrogen bonds	OH, NH groups	Moderately strong
Ionic attractions	Ionizable or charged groups	Strong

The term HETP (height equivalent theoretical plate) is often used and it defines the size of the theoretical plate (section 5.2.2). The smaller the size, the more theoretical plates occur in a column of unit length and this results in a better separation.

The affinity of a compound for any phase depends on the attractive forces between the molecules of the compound and those of the phase. These forces are of several types, shown in Table 5.1, and are influenced by the structure of the molecule and its functional groups. The intermolecular forces are also influenced by the shape and size of the molecules. The type of force that plays the greatest part in the interaction varies considerably with the form of chromatography that is being employed and will be discussed in more detail later.

5.2.2 The concept of the HETP

The actual process of separation of two compounds is conceived as a series of 'quantum leaps', one for each HETP, as described and illustrated in Figure 5.1, although, in reality, the mobile phase flows evenly along the particular path. Each 'quantum' is the HETP which corresponds to a small volume containing both phases where equilibrium takes place. However, it must be stressed that the idea of HETP is very much a theoretical concept.

Because of their different distribution coefficients, the two substances are differently distributed at equilibrium in the first HETP. The mobile phase then carries its components to the next HETP where the dynamic equilibrium is again achieved. At the same time, a fresh aliquot of mobile phase has moved into the 'first' HETP and the substances distributed in the stationary phase redistribute themselves between the two phases. The units of mobile phase then move on one more HETP and the process is repeated. After many such events, the substances separate into discrete zones in the mobile phase and these move further apart from each other as they progress through the system. In other words, the zones are carried along by the mobile phase at different velocities, the highest velocity

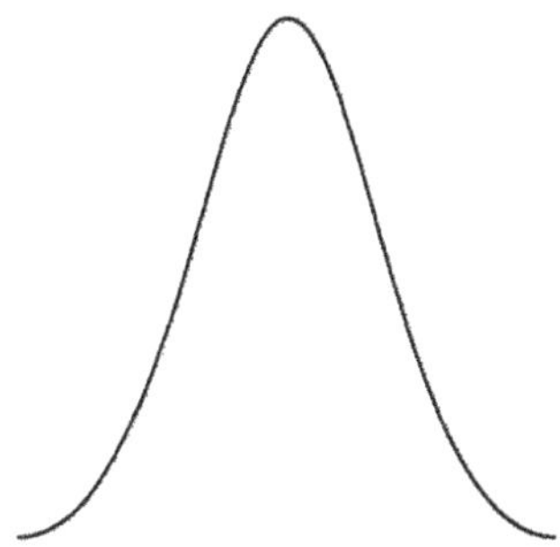

Figure 5.2 Gaussian peak for a substance separated by a chromatographic process.

being demonstrated by the zone which has the least affinity for the stationary phase, until the end of the chromatographic process.

Chromatographic experiments either measure the volume of mobile phase (often expressed as time if the mobile phase is moving at constant velocity) each zone needs to reach the end of the stationary phase, or measure the distance moved by each zone in a chromatographic system after the mobile phase has moved a fixed distance. The chromatograms produced therefore consist of either a series of peaks, where the horizontal axis is the elution time, or a series of zones at different distances from the starting point.

Each zone contains a compound theoretically spread in a Gaussian distribution, and this is often seen as a characteristic curve on a chromatogram, known as a chromatographic peak (Figure 5.2). The shape displayed can be proved mathematically by considering the processes taking place. In most situations the production of a sharp, narrow peak is the aim. The shape of the peak depends on a variety of factors (discussed below) and is one of the two major factors influencing resolution.

5.2.3 Chromatographic resolution

In most chromatographic separations it is necessary to achieve good resolution. Resolution is the assessment of the separation of two peaks or zones and depends on two factors: the distance between their points of maximum concentration and the width of each zone. It is given by the formula:

$$R = \frac{R_a - R_b}{(w_a - w_b)/2}$$

Illustrations of these terms are shown in Figures 5.3 and 5.4.

The distance between the two zones ($R_a - R_b$) is an expression of the column selectivity and is dependent on the relative distribution

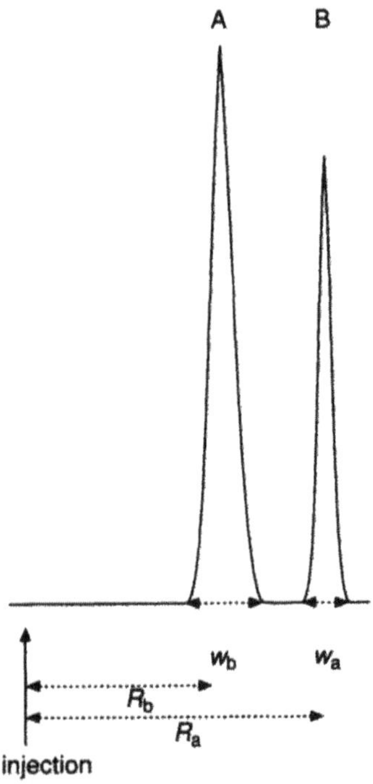

Figure 5.3 Resolution of chromatographic peaks.

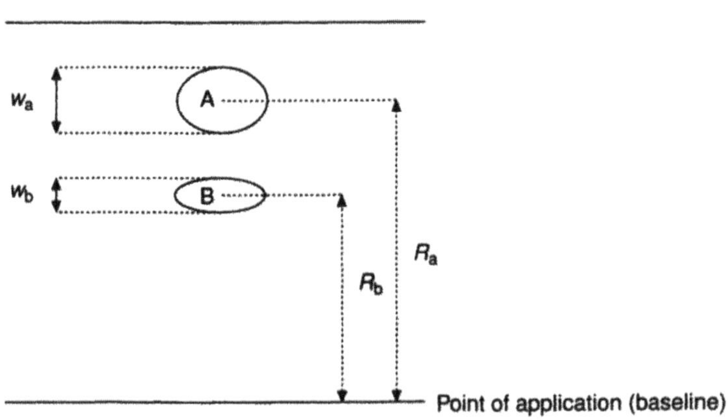

Figure 5.4 Resolution of zones on a layer chromatogram.

coefficients of the components; the larger the difference, the further apart will be the two zones. It can be changed by altering one or both of the phases. The other aspect of resolution, the column efficiency, is associated with the width of each peak $w_a - w_b$. From the equation above it can be seen that for good resolution this term should be small, i.e. a sharp peak is produced indicating that the substance occupies only a small volume of the mobile phase at the end of the chromatographic separation.

Good column efficiency depends on:

- a small particle size of the stationary phase or its support;
- regular packing of the stationary phase (which should be of a narrow size range);
- a thin layer of stationary phase;
- an optimum flow rate and temperature.

All these factors can be influenced by the choice of materials and the manipulation of a chromatographic experiment, and it is the knowledge of how these somewhat theoretical concepts can be applied in practical situations that can achieve good chromatographic results, particularly in the separation of distinct fractions from a mixture (section 7.1.1).

The width of the zone is also dependent on the size of the initial application to the chromatographic system, and much of the skill in achieving good chromatographic fractionation involves the ability to apply the sample mixture properly so that it occupies as small a volume at the start of the system as possible and reduces band broadening due to diffusion. This latter effect is also minimized by reducing the time during which chromatographic separations occur.

5.3 PHYSICO-CHEMICAL PROCESSES TAKING PLACE IN CHROMATOGRAPHY

The distribution between the two phases, which is an essential feature of chromatography, can be one or more of a range of physico-chemical phenomena, although usually one is predominant for each technique. The particular distribution system is very dependent on the type of molecule being separated and the chromatographic materials used. The major types of materials connected with each phenomenon are shown in Table 5.2.

NB Often a mixture of the processes described below takes place and Table 5.2 oversimplifies the real-life situation, e.g. moisture within silica gel can introduce a partition process in addition to the adsorption process which is the major one influencing separation. The major distribution processes are discussed below.

5.3.1 Adsorption

Adsorption is the distribution of a substance between the surface of a solid and a liquid in which it dissolves (Figure 5.5). It is considered that there is a finite number of sites on the surface of the solid to which the molecules are attached. Therefore, if the solution is too concentrated, the

Table 5.2 Summary of chromatographic processes and techniques

Process	*Chromatographic technique*	*Typical stationary phases*	*Mobile phase*
Adsorption	Column, thin-layer (TLC), some HPLC	Silica gel (dry), alumina, polyamide (H bonding)	Non-polar or wet liquids
Partition	Paper, TLC, liquid–liquid, gas chromatography (GC)	Paper, cellulose, specialized phases for GC	Polar or moderate liquids, gas
Reversed-phase partition	High-performance liquid chromatography (HPLC), impregnated layers	Bonded silicas, paraffin, dimethylformamide, graphite carbon	Polar or moderate liquids
Ion-exchange	Column, HPLC	Specialized resins	Aqueous
Size exclusion	Column, HPLC	Specialized polymers, e.g. Sephadex™	Aqueous, moderate polar organic liquids
Affinity	Column	Specialized gels with attached ligands	Aqueous

constant equilibrium cannot occur because all the sites are occupied and the concentration in the liquid phase is too high. If this applies, the system is said to be overloaded and the chromatographic separation is not very good. It is important to remember this when deciding how much of a mixture to apply to any chromatographic system relying chiefly on adsorption. The ratio of distribution between the two phases after equilibrium has been reached is called the adsorption coefficient.

The adsorption coefficient varies with temperature, usually decreasing with a rise in temperature, i.e. relatively less substance is adsorbed as the temperature increases. In chromatographic terms this means that the retention time (section 8.2.2) will decrease and the R_f value (section 8.2.2) will increase as the temperature rises.

Adsorption is the major process causing separation in conventional thin-layer chromatography and in conventional column chromatography. The adsorbent consists of fine particles forming a powder which is spread as a layer or is packed into a tube.

Common adsorbents are silica gel and alumina but charcoal, kieselguhr and other finely divided solids are sometimes used. Conventionally, adsorbents such as silica gel, alumina and kieselguhr are more polar than the mobile phase, due to the many OH groups on their surfaces, but charcoal is a good example of a non-polar adsorbent since it presents a surface of covalently bonded carbon atoms with no polar groups.

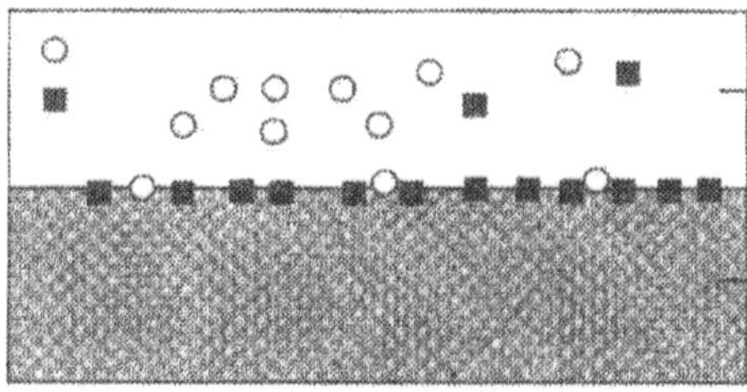

Figure 5.5 The process of adsorption; ■, molecules with adsorption coefficient 4 : 1; ○, molecules with adsorption coefficient 1 : 3.

5.3.2 Partition

When a substance distributes itself in solution between two immiscible fluids an equilibrium is reached and the ratio of the concentrations is a constant, the partition coefficient (Figure 5.1). The fluids are either liquid or gas: the stationary phase is always a liquid and the mobile phase either a liquid or a gas. A specialized form of partition chromatography uses the supercritical fluid form, which combines some properties of gases with those of liquids, but this is not very applicable to most biological materials for fractionation. The stationary-phase liquid has to be immobilized in some way so that the mobile phase can pass across it and have a large area of interface.

The earliest forms of partition chromatography made use of the water immobilized within the cellulose fibres in paper as the stationary phase and this has been extended to the use of more refined forms of cellulose. Immobilization can also be achieved by impregnating porous substances, such as kieselguhr, with a viscous liquid so that it is coated as a fine film over a very large surface area. In gas chromatography many specialized substances have been developed to provide viscous, non-volatile, thermolabile phases with a range of polarities (Table 8.1) and these are either coated on such an inert support or, in capillary chromatography, on the inside wall of very small-diameter tubes. Another method of immobilization of the stationary phase has developed because of high-performance liquid chromatography (HPLC) and comprises the chemical bonding between a fairly substantial molecule and the surface functional groups of a high surface area solid. The bonded molecules form a monomolecular layer that functions as a liquid as far as partition of the solutes is concerned. A major example of this is the use of octadecyl silica (ODS) which is widely used in liquid chromatography and has the structure shown in Figure 5.6.

Bulk liquid–liquid systems are known as counter-current chromatography and the stationary phase is immobilized in various ways depending on the design of the apparatus (section 6.2).

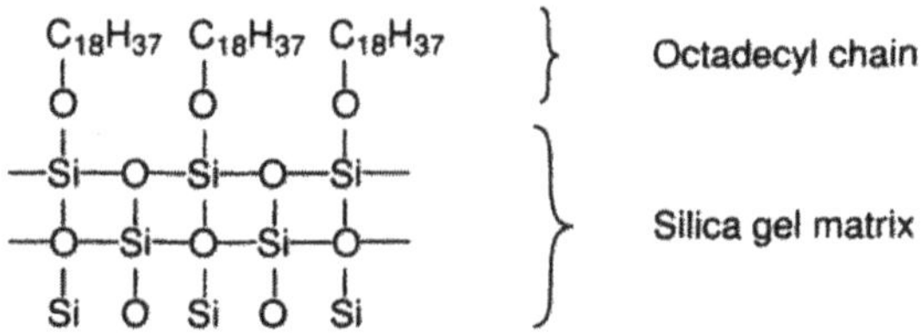

Figure 5.6 Structure of ODS reversed-phase silica.

The solubility of a substance in each phase is related to the attraction between the molecules of the phase and the solute molecules. Polarity and ionizable groups are the major factors influencing such attraction. In the latter case, changes in pH of the system can make great differences in the separation achieved.

The partition coefficient varies with temperature. As the temperature increases the partition coefficient tends to increase in favour of the phase in which the solutes are less soluble. This is usually the mobile phase, so an increase in temperature results in a decrease in retention time (or increase in R_f values on layers). This is particularly noticeable in gas chromatography, where retention times decrease dramatically with relatively small increases in temperature.

Conventionally, the stationary phase is more polar than the mobile phase and is known as normal-phase chromatography. When the opposite applies the system is known as reversed-phase chromatography, often designated RP. In conventional partition chromatography the least polar substances have the shortest retention time, whereas in RP chromatography the solutes are eluted in order of decreasing polarity. Such systems, where the stationary phase is less polar than the mobile phase, are now used extensively in analytical HPLC for separation of aqueous extracts, following the development of bonded reverse-phase silicas and other materials. The OH groups on the surface of the silica are treated with reagents to produce bonded structures, e.g. long hydrocarbon chains such as octadecane, linked by an ether bridge (Figure 5.6). Many different types of compounds have now been bonded by this method and a wide range of RP materials are available commercially. Although originally designed for analytical purposes, these RP stationary phases are now being used increasingly in a preparative context for the fractionation of polar extracts. A cheaper form of reversed-phase stationary phases is purified charcoal powder; however, it is messy and irreversibly binds some compounds. More sophisticated purified graphite carbon columns are used in HPLC.

5.3.3 Ion-exchange

Ion-exchange chromatography could be viewed as a type of adsorption process. The stationary phase consists of a porous matrix, usually called a 'resin', through which the mobile phase can pass. The matrix consists of a cross-linked polymer to which groups carrying a positive or negative electrical charge, the ligands, are attached.

Substances with the opposite electrical charge dissolved in the mobile phase (usually an aqueous buffer solution) will be attracted to the charged groups on the stationary phase and an equilibrium will be established between the concentrations of such molecules between the two phases. The affinity of any substance to the stationary phase will be related to several factors, such as the ionic strength and pH of the mobile phase, the temperature and the type of group bearing the opposite charge.

Ion-exchange chromatography can be used in fractionation in two separate ways: either to separate all charged ions of one type from a mixture or to separate the different types of molecule bearing a particular charge from each other.

Separation of molecules with a common charge is possible when the equilibrium in a particular environment is very much in favour of the molecules being attached to the stationary phase. Practically all of these charged molecules become bound to the ligand, and thus the stationary phase, and the flow of the mobile phase carries all the other components through so that they elute from the chromatographic system. The bound molecules are then released from the stationary phase by altering the conditions, e.g. changing the pH of the mobile phase or the concentration or type of other ions present. This release is brought about because the equilibrium shifts much more in favour of the mobile phase (e.g. by the sample molecules becoming non-ionized) or because other ions, called counter-ions, displace those of interest. This process is shown in Figure 5.7.

The alternative approach of ion-exchange fractionation (separation of similarly charged compounds) is more akin to other forms of chromatography. Although they may all carry the same type of charge which attracts them to a ligand, different substances will have different affinities for it, depending on several factors such as molecular size and shape, number of ionizable groups and pK_a value. These different affinities result in different retention times and so compounds from a mixture are separated when it is passed through a column. This process is shown in Figure 5.8.

Most ion-exchange chromatography, particularly on the scale used for fractionation, uses stationary phases which, conventionally, consist of beads of resin consisting of various types of polymers to which the ligands (charged groups) are attached by chemical bonds. In recent years ionic

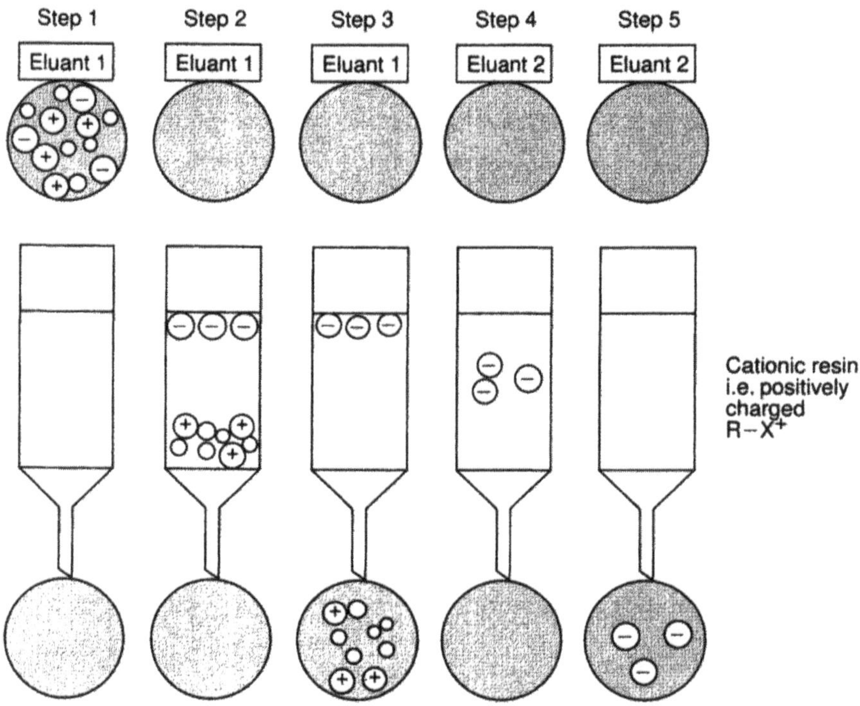

Figure 5.7 Ion-exchange process; ⊕, molecules carrying a positive charge; ⊖, molecules carrying a negative charge; O, molecules carrying no charge. Step 1, the extract containing a mixture of molecules is dissolved in eluting solvent 1 and introduced to the column. Step 2, negatively charged molecules are retained by the cationic resin. Step 3, as elution with solvent 1 continues, the neutral and negatively charged molecules are eluted. Step 4, a new eluting solvent is introduced which contains anions which have a higher affinity for the stationary phase than the anions in the mixture. The sample anions are displaced and start to move through the column. Step 5, the sample anionic molecules are eluted.

exchangers bonded to silica have been introduced for column chromatography or attached to cellulose for ion-exchange paper chromatography.

Where resins are concerned, the beads have to be left in the mobile phase for several hours to allow for swelling to occur prior to use. Styrene and divinylbenzene polymers are the most common types encountered. Positively charged groups in cationic resins, or anion-exchangers, are usually based on amines. Quaternary ammonium attachments to the polymer chain ($-R_3N^+$) are strong bases which are ionized over a wide pH range and give strong exchangers, whereas tertiary ($-R_2N$) and secondary amines ($-RNH^+$) protonate and become basic in nature over a shorter pH range and are therefore called weak exchangers. (NB It is important to recognize the potential confusion in terminology between the description of the resin and its exchange species.)

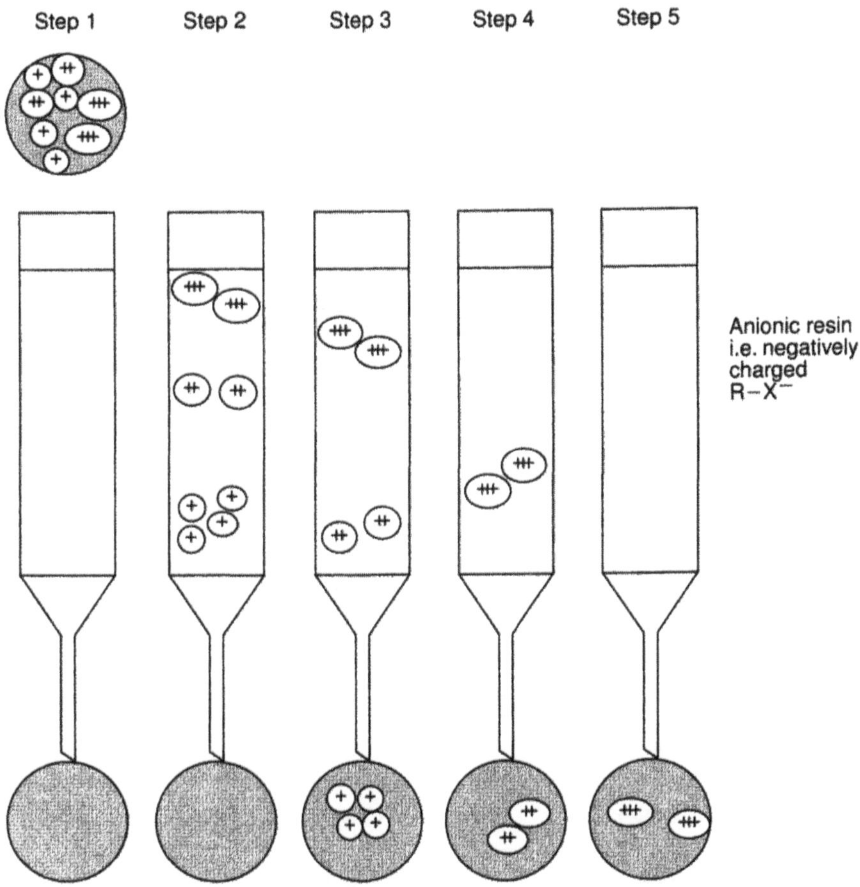

Figure 5.8 Ion-exchange separation of molecules bearing the same charge; ⊕ (++) (+++), molecules in order of increasing positive charge. Step 1, extract containing a mixture of cationic molecules is introduced to the column. Step 2, the molecules with the smallest charge move most quickly through the column. Step 3, the molecules with the smallest charge are eluted from the column. Steps 4, 5, molecules are eluted from the column in bands, those with the highest net charge being eluted last.

Anionic resins (i.e. cation-exchangers) have acidic groups attached to the polymeric chain. Strong acid groups are based on sulphonic acids ($-SO_3H$) which carry a negative charge in most pH conditions, whereas more weakly acidic ligands are comprised of carboxylic acid groups ($-COOH$). These are only charged in conditions where the pH is fairly high (>5).

Other types of support for the charged groups are sometimes encountered, particularly modified cellulose in the form of paper or as a powder. Silica-based materials are available but have the disadvantage of being unstable outside a pH range of 3–6. A selection of ion-exchange stationary phases is shown in Table 5.3.

Table 5.3 A selection of ion-exchange stationary phases

Type of exchange resin	*Name*	*Chemical group*	*Special conditions*
Weak cationic (i.e. anion exchange)	Amberlite IRA-93	Polyamine on polystyrene	pH 0–7
	Dowex MWA-1	Polyamine on polystyrene	pH 0–7
	Duolite A-7	Amine on polystyrene	pH 0–6
	DEAE Sephadex	Diethylaminoethyl dextran	For molecules $<3 \times 10^4$ daltons
	DEAE Sephacel	Diethylaminoethyl cellulose	For molecules $<1 \times 10^6$ daltons
Strong cationic (i.e. anion exchange)	Amberlite IRA-400	Quaternary ammonium on polystyrene	pH 0–14
	Dowex 1X2-400	Trimethylbenzyl-ammonium on polystyrene	pH 0–14
Weak anionic (i.e. cation exchange)	Amberlite IRC-50	Carboxylic acid on polystyrene	pH 5–14
	Diaion WK 20	Carboxylic acid on polystyrene	pH 4–14
	CM Sephadex	Carboxymethyl on dextran	For molecules $<3 \times 10^4$ daltons
	CM Cellulose	Carboxymethyl cellulose	
	Duolite C-433	Polyacrylic acid	pH 5–14
Strong anionic (i.e. cation exchange)	Amberlite 120	Sulphonic acid on polystyrene	pH 0–14
	Diaion SK 1B	Sulphonic acid on polystyrene	pH 0–14
	Dowex −50W	Sulphonic acid on polystyrene	pH 0–14

5.3.4 Size exclusion

Size exclusion chromatography is also known as gel filtration or gel permeation chromatography. The principle of this technique is a little different from the others described here since the two phases are distinguished only by their immediate environment. The stationary phase consists of the fluid contained within a porous three-dimensional polymeric matrix

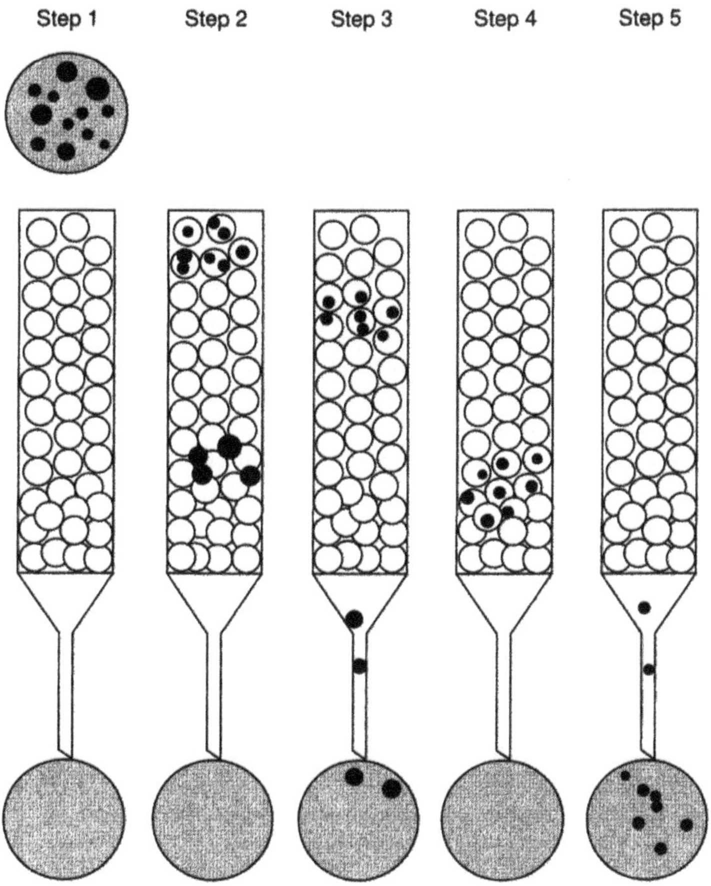

Figure 5.9 Exclusion chromatography process; ●, molecules larger than pore size of gel; •, molecules smaller than pore size of gel. The same eluant is employed throughout the separation. Step 1, extract containing a mixture of molecules of different size is introduced to the column. Step 2, the largest molecules move most quickly through the column, since they cannot enter the pores of the gel. Step 3, the largest molecules are eluted, the smaller molecules keep moving through the gel but move less quickly since they enter the pores of the gel. Steps 4, 5, the smaller molecules are eventually eluted from the gel.

with pores not greater than a certain size. These pores result from regular crosslinking formed by chemical reaction between long-chain polymers and bridging reagent, and the size of the pores can be diminished by increasing the number of crosslinks. This is achieved during manufacture of the gel by increasing the proportion of crosslinking agent. The mobile phase is the fluid outside the gel matrix.

Separation occurs based on the different sizes, and to some extent, the different shapes of the molecules dissolved in the mobile phase and this is only really applicable to molecules of greater than 300 daltons. Large molecules are unable to enter the pores, and so are unable to enter the stationary phase, and pass through the chromatographic system very quickly (Figure 5.9).

Smaller molecules can enter the pores and penetrate the mobile phase. Since the pores cover a range of sizes below the maximum, the smallest molecules enter more pores than those of larger size. Therefore the equilibrium is more in favour of the stationary phase for smaller and more spherical molecules, with the consequence that these types of compound have the longest retention times.

Fractionation using size exclusion chromatography is most frequently used to separate small molecules from larger ones of interest such as polysaccharides or proteins. In this case the fractions first eluted from the column are of interest. Alternatively, high molecular mass compounds may be viewed as unwanted material where attention is directed towards the low molecular mass components of a mixture, and so the later fractions are required in this case.

The original material developed for exclusion chromatography was called Sephadex™ and consisted of a dextran gel (polymeric 1–6 glucose crosslinked using epichlorhydrin). A range of grades with different pore sizes was available but their use was limited to aqueous solutions. Many different types of matrix have now been developed using different types of polymer so that relatively non-polar liquid mobile phases can be used. Gels similar to those used in ion exchange (but containing no charged ligands) and alkylated dextran gels, e.g. Sephadex LH-20™, are frequently used for this purpose. A selection of gels commonly used is shown in Table 5.4.

Table 5.4 A selection of exclusion chromatography stationary phases

Name of gel	*Type of gel*	*Special conditions*	*Fractionation range (in daltons × 10^3)*
Sephadex G-10	Dextran crosslinked with epichlorhydrin	Water-soluble compounds	<0.7
Sephadex G-25	Dextran crosslinked with epichlorhydrin	Water-soluble compounds	0.1–5.0
Sephadex G-100	Dextran crosslinked with epichlorhydrin	Water-soluble compounds	1–150
PDX GF-25	Crosslinked dextran	Water-soluble compounds	1.0–5.0
Beaded cellulose		Water-soluble compounds	100–3000
Sephadex LH-20	Hydroxypropyl-dextran	Less polar compounds	<1.5

The gels are available commercially in a bead form which has to be immersed in the mobile phase for several hours before use, to allow swelling to take place. Batches of gel can be left swollen for some time so that they can be used when required. (NB Dextran-based gels are susceptible to fungal and bacterial attack. A small amount of preservative, such as sodium azide 0.1% w/v, is often added to the swelling medium to prevent this. This must be washed out with eluant before the column is used for fractionation.)

5.3.5 Biochemical affinity

Affinity chromatography is based on biochemical interactions between the stationary phase and solutes in the mobile phase. These processes include antigen–antibody, enzyme–substrate and drug–receptor interactions. Affinity chromatography is primarily used to separate a group of substances with a high affinity for the stationary phase from all others present in an extract or solution. The separated compounds can then be eluted from the stationary phase by changing the prevailing conditions.

The structure of the stationary phase in affinity chromatography is a matrix very similar to that described above for ion-exchange chromatography, except that the ligands in this case consist of a suitable substrate, antigen, etc. bonded on to the matrix. The sample is usually eluted off the column by passing an eluant containing dissolved ligand through the column. Separation of the sample–ligand complex can be achieved by altering the pH or polarity of the eluant or by dialysis if appropriate (section 4.3.4). In affinity chromatography the matrix usually has very large pores and, since aqueous extracts are the most common type of mixture to be fractionated, a hydrophilic medium such as agarose gel is employed.

Customized gels can be purchased or prepared for particular situations and for the separation of specific types of substance. Affinity columns with sugar ligands can be used for the isolation of plant and animal lectins from crude extracts.

5.4 BIBLIOGRAPHY

Braithwaite, A. and Smith, F.J. (1985) *Chromatographic Methods*, 4th edn, Chapman & Hall, London.

Fifield, F.W. and Kealey, D. (1990) *Principles and Practice of Analytical Chemistry*, 3rd edn, Blackie, Glasgow.

Hostettmann, K., Hostettmann, M. and Marston, A. (1986) *Preparative Chromatography Techniques*, Springer, Berlin.

Ikan, R. (1991) *Natural Products – A Laboratory Guide*, 2nd edn, Academic Press, London.

Stahl, E. (1969) *Thin Layer Chromatography*, 2nd edn, George Allen and Unwin, London; Springer, Berlin.

CHAPTER SIX

Chromatographic procedures using columns and liquid–liquid systems

There are two general types of chromatographic techniques, planar and column systems. In the first of these the separation is usually terminated after a fixed time and the different distances moved by each zone are measured. In column techniques the path length for each compound in a mixture is the same (i.e. the full length of the column) and it is the time taken to travel this distance (or the volume eluted under constant flow conditions) which is measured. However, it should be noted that when preparative chromatography is employed the elution of fixed-volume fractions in numerical sequence is most often used and little attention is paid to the actual time involved.

6.1 COLUMN CHROMATOGRAPHY

Column chromatography is the oldest form of chromatographic technique. A tube is packed with a solid stationary phase, the sample mixture applied to the top of the column and the mobile phase is allowed to move down through the column under gravity. The constituents of the mixture move at different rates through the column as bands. Eventually they reach the end of the column and elute in solution in the mobile phase (Figure 6.1). Practical details for packing and running columns are given in section 9.3 and the steps involved in setting up a column are described in section 9.3.1.

6.1.1 Conventional column chromatography

Conventionally, the stationary phases in column chromatography were adsorbents of high polarity, e.g. silica gel. However, recent years have seen the introduction of many other forms of stationary phase, e.g. reverse-phase silica, ion-exchange resins, exclusion chromatography stationary phases.

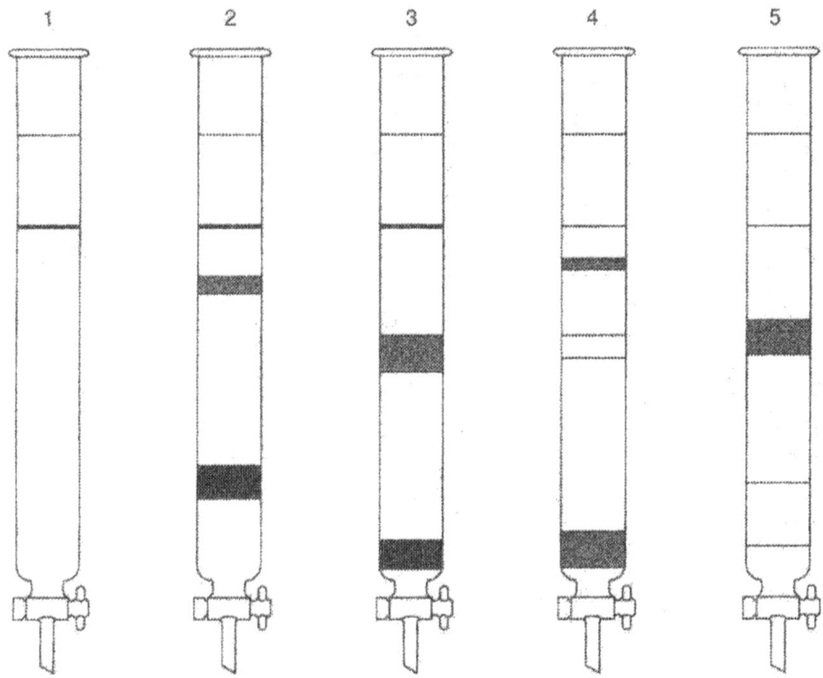

Figure 6.1 Separation on a chromatography column. 1, Elution commences with mobile phase A. 2, Elution continues with mobile phase A. 3, Elution continues with mobile phase A, first compounds elute. 4, Elution commences with mobile phase B, baseline materials start to move through column. 5, Elution continues with mobile phase B, eventually all bands elute from the column.

The mobile phase passes through the column under the force of gravity and, due to the resistance of the stationary phase packed in the column, the flow rate is not very high. Elution with the mobile phase can be isocratic, where mobile phase of only one composition is used, or gradient. Gradient elution consists of a sequence of different compositions of the mobile phase, most commonly as a gradient of increasing (or, in the case of reverse-phase systems, decreasing) polarity. The gradient is achieved most commonly as a series of 'steps' consisting of aliquots of mobile phase of fixed composition. More sophisticated systems are now available where a linear gradient is established by means of a mixing pump which uses two separate reservoirs to deliver a constantly increasing concentration of one component in the eluting solvent.

The bands of compounds which eventually elute from the column cannot be seen unless coloured substances are being separated. The usual method of detecting the zones of eluted compounds is to collect the eluate from the column, usually performed automatically using a fraction collector, as consecutive fractions measured by a fixed time or volume.

These fractions can then be analysed either biologically (by a bioassay), chemically (for the presence of compounds of interest by further chromatography, particularly TLC) or physically, e.g. absorbance. These topics are dealt with in greater detail in Chapter 8.

A major drawback of conventional column chromatography using solid packed absorbents is the time involved because the mobile phase moves very slowly. The bands therefore broaden by diffusion as they move through the column and are not well resolved from one another. (NB There is an optimum speed for elution of a column to achieve good resolution. If the mobile phase flow rate is too great, equilibrium cannot occur and separation of components does not take place.) A long separation time is also disadvantageous since compounds are more likely to decompose or diffuse through the mobile phase if left on the column for a considerable period of time. Originally, resolution was also poor due to the large range of particle size in the rather crude stationary phases used (40–100 μm). Stationary phases with smaller particle sizes were developed but reduced the solvent flow to unacceptable levels if gravity only was the driving force.

6.1.2 Practical steps in conventional column chromatography

Choice of stationary and mobile phases

The stationary phase to be used is largely dependent on the types of molecules that are present, as discussed in section 8.2. The choice of mobile phase for adsorption and partition column chromatography can be made by TLC analysis. The methods of selecting a TLC system which gives good resolution are described in section 7.1.1.

Adsorption chromatography

When an adsorption process is exploited in a column, it must be realized that TLC is carried out using layers that have been dried well, but, in contrast, the same substance used as a stationary phase in a column very often contains appreciably more water. The water present occupies a number of the adsorption sites and so reduces the efficiency of the adsorbent and decreases the retention of compounds, leading to decreased elution times.

To overcome this effect, it is a useful practice to reduce the polarity of the mobile phase mixture for normal phase separation giving best results on TLC by increasing the proportion of the least polar component by a factor of about two, e.g. if TLC on silica gel gave good resolution of a mixture with chloroform : methanol 6 : 1, then the initial mobile phase used in the column should be chloroform : methanol 12 : 1.

It is usual to apply gradient elution to a column so, for the example given, the initial eluant would be chloroform : methanol 12 : 1 followed by chloroform : methanol 9 : 1, 6 : 1, 3 : 1, etc.

Partition column chromatography

The general method for determining a suitable system based on partition is much the same as that described above for adsorption, although the nature of the two major types of stationary phases used influences the range of mobile phases tested.

The stationary phases used in partition chromatography consist of two types, the polar ones, such as water, held within a matrix such as cellulose, and the non-polar monomolecular layer reverse-phase type, such as octadecyl chains bound to silica gel (section 5.3.2).

Thin-layer chromatographic plates are available that contain these types as the stationary phase and so can be used to test the suitability of mobile phases for use in a column using the same material as the stationary phase. In the case of the first type, chromatographic paper can be used instead of cellulose thin layer. The mixture is applied and run in a series of mobile phases of different polarities.

Where water held by cellulose is the stationary phase, the mobile phases are normally quite polar and should consist of water mixed with less polar but miscible substances, such as acetic acid, butanol or phenol.

In the case of reverse-phase layers the range of mobile phases ranges in polarity from water to acetone. Alteration of the pH of the mobile phase can have a great effect on the separation. If ionization of the mixture components may occur, a small amount of a salt such as ammonium nitrate may be added to the water to help in the formation of 'ion pairs' which function as less-polar entities than the unpaired ions.

Since partition separations are less sensitive to the presence of small quantities of water, no adjustment normally needs to be made when the system is adapted from layer to column separation.

Ion-exchange chromatography

Aqueous solutions are usually employed as mobile phases in ion-exchange chromatography. The choice of stationary phase depends on the type of ions to be separated as the phase should bear the opposite charge to these components. There is no easy way to test a system, although ion-exchange papers, which have ionizable groups chemically bound to the cellulose molecules in the paper, are commercially available and can be used for initial determinations.

The nature of the mobile phase can vary according to the pH, usually through using different buffer solutions, and also according to the ionic strength, particularly of ions that might displace the analyte ions from their sites on the stationary phase.

In most fractionation procedures, ion exchange is used to separate charged from non-charged components and so these finer points, which affect the resolution of the different species of charged analytes from each other, are not usually relevant.

An exercise involving ion-exchange chromatography is given in section 10.2.9.

Exclusion chromatography
The choice in exclusion chromatography is mainly in the type of stationary phase to be used. In all circumstances the gel must be allowed to swell in the mobile phase for some time before it is packed into the column.

Many separations involving exclusion chromatography have a very hydrophilic gel stationary phase and aqueous mobile phases. The stationary phases available vary according to the size of the pores, so it is necessary to have some idea of the range of molecular sizes present in the mixtures in order to choose the best one for the situation. Layers of exclusion phases are available but they are costly, rather fragile and not so versatile as conventional thin layers, so they are seldom used for method development. A small column can be packed and tested with a little of the sample and eluted in order to assess the suitability of a proposed method.

An exercise involving exclusion chromatography is given in section 10.2.4.

6.1.3 New column chromatography procedures

Improvements in resolution in column chromatography have been made by speeding up the passage of the mobile phase and by the introduction of better commercial stationary phases which have a reduced particle size and range of sizes (10–20 μm).

The time for separation has been reduced by the introduction of positive or negative pressure on the mobile phase to force it to run through the stationary phase bed more quickly. These techniques are known as flash chromatography and vacuum liquid chromatography, respectively. Increased pressure also makes it possible to use a stationary phase of smaller particle size, which improves resolution. This is further enhanced by the commercial availability of more refined stationary phases with a narrow range of particle sizes specifically designed for the column procedures described below.

Flash chromatography

Flash chromatography is simply column chromatography where the mobile phase passes through relatively quickly because of the application of positive pressure, usually about 10 bar, from a nitrogen cylinder or from an on-line supply. Compressed air could be used but, since this contains oxygen and water vapour, it is more likely to lead to decomposition of the products present in the extract and to changes in the chromato-

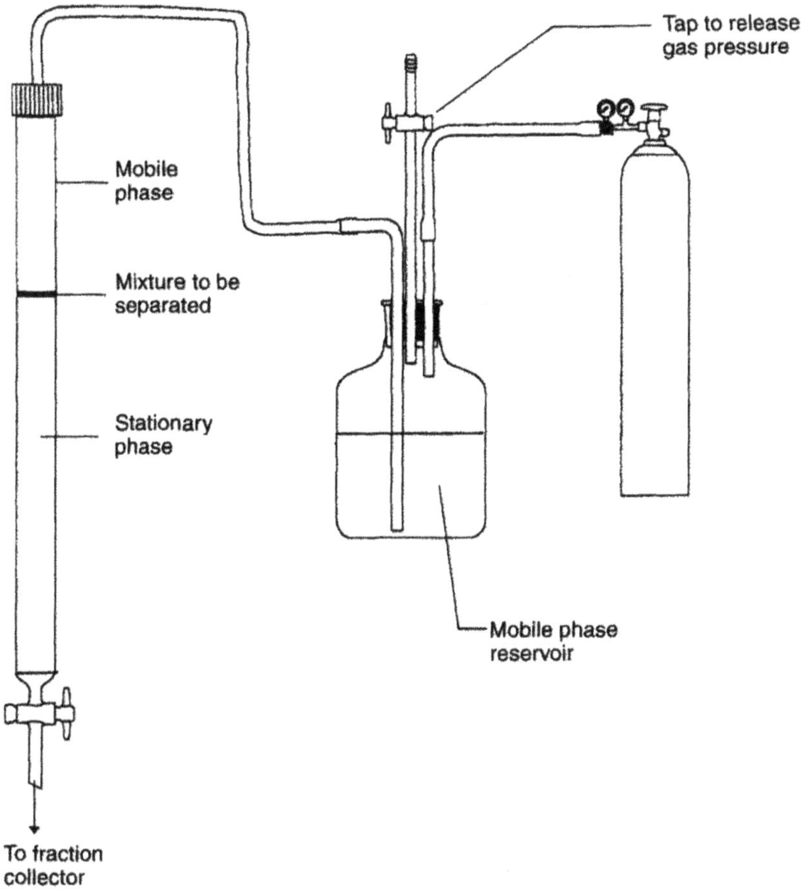

Figure 6.2 Flash chromatography column apparatus.

graphic separation. A range of different, small particle size, narrow size range adsorbents designed for flash chromatography is now available from manufacturers of chromatographic materials.

Pressure is applied to the mobile phase either directly after it is introduced into the column or indirectly to an enclosed reservoir connected by tube to the column, as shown in Figure 6.2. (NB Flash chromatography can only be carried out effectively with a stationary phase of the correct particle size. Some materials, particularly TLC-grade silica, have particles which are too fine for flash chromatography but can be used for the vacuum liquid chromatography technique described below.)

An exercise involving flash chromatography is given in section 10.2.6.

Vacuum liquid chromatography

Vacuum liquid chromatography (VLC) is a form of column chromatography that is especially useful for the rapid crude fractionation of extracts. The 'column', which may be either normal or reverse-phase TLC-grade adsorbent, is relatively squat and the mobile phase is sucked through by reduced pressure. Fractions are usually collected as aliquots of eluants of one polarity. The sequence of eluant aliquots can be designed to produce a stepwise gradient elution.

The 'column' is usually contained in a Buchner funnel and the eluant collected in a Buchner flask. However, this arrangement means that the flask has to be disconnected after each elution. A useful modification is to attach a separating funnel connected to the Buchner funnel via a T-piece adaptor, with the side-arm connected to the vacuum supply, as shown in Figure 6.3. This allows the eluant to be drained off from the collecting separating funnel (as long as the vacuum is released) without any dissembling of the apparatus.

Exercises involving VLC are given in sections 10.2.5 and 10.2.7.

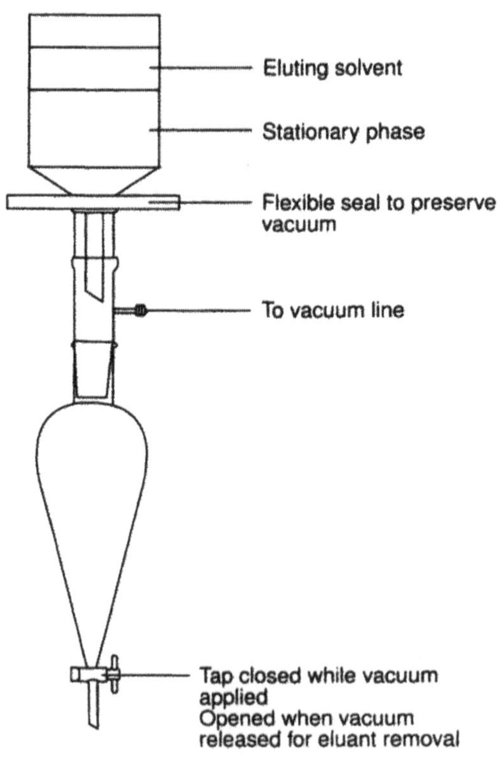

Figure 6.3 Vacuum liquid chromatography apparatus.

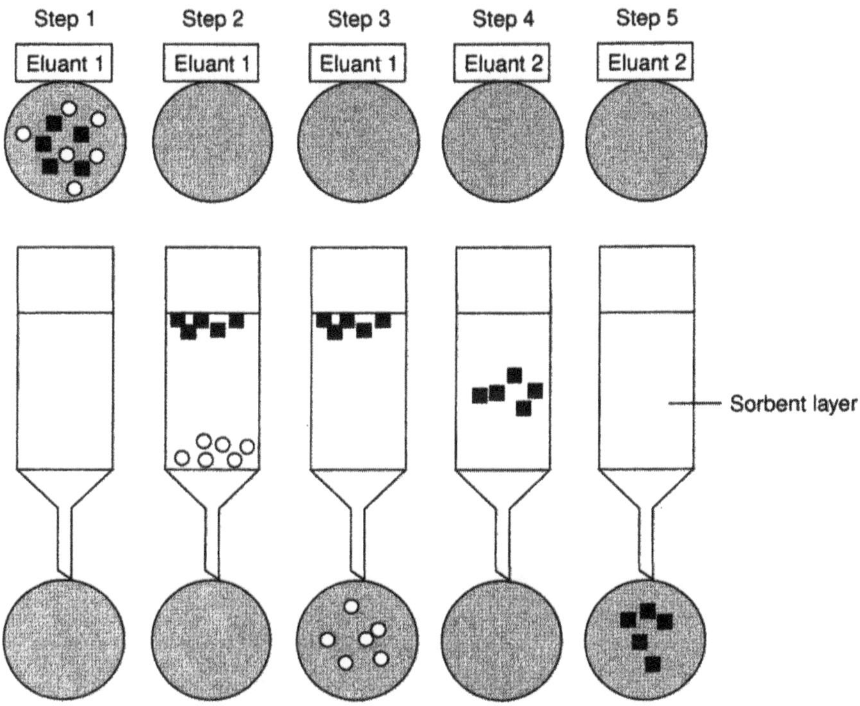

Figure 6.4 Sorbent extraction process; ■, required molecules; ○, molecules not required. Step 1, extract containing a mixture of molecules dissolved in solvent 1 is introduced into the column. Step 2, the required molecules are retained by the sorbent (solid phase). Step 3, as elution with solvent 1 continues, the unwanted molecules are eluted. Step 4, a new eluting solvent is introduced for which the required molecules have a high affinity. The required molecules start to move through the column. Step 5, the required molecules are eluted.

Sorbent extraction

Sorbent extraction is a form of column chromatography in which the compounds of interest are removed from solution by adsorption on to the stationary phase rather than being eluted in the mobile phase. After passage of the original mobile phase, the compounds of interest are adsorbed on the stationary phase whence they are removed by eluting with a suitable solvent – either with a different polarity than the initial solvent or by a change in pH. The process is shown in Figure 6.4.

Sorbent extraction can be used with any type of stationary phase except those used for exclusion chromatography.

An exercise involving sorbent extraction is given in section 10.2.5.

Large-scale adaptations of HPLC

HPLC using a closed, prepacked column can be used preparatively if the pump used to deliver the mobile phase has sufficient capacity and power to force it through the column. The technique may be called medium-pressure liquid chromatography if the stationary phase particle size is larger than that used for analytical HPLC.

These methods are expensive but have the advantage that the extract can be injected on-line so that the initial application is quite small. Commercial prepacking of the column also ensures a good resolution of the components.

Preparative adaptations of HPLC are being used increasingly in laboratories where the isolation of natural compounds is performed regularly, but their cost prohibits casual use.

6.2 LIQUID–LIQUID PARTITION SCHEMES

6.2.1 Introduction

There is some debate over whether these are really chromatographic systems but they fulfil all the criteria for such. They are sometimes known as countercurrent distribution systems and rely, to an almost exclusive extent, on exploiting the different partition coefficients of the components of a mixture to achieve separation.

It is important to realize that many forms of column chromatography with solid stationary phases such as cellulose or reversed-phase silica also rely mainly on partition phenomena. The major difference between these techniques and liquid–liquid systems is that the latter involve no solid material at all. The advantage of liquid–liquid systems is that, theoretically at least, no material is lost through irreversible binding to solid matter.

6.2.2 Simple methods

The simplest form of liquid–liquid chromatography is a series of separating funnels, each containing the stationary phase. The mixture, dissolved in the mobile phase, is introduced to the first funnel and the mixture shaken carefully but thoroughly so that equilibrium is reached. The mobile phase is then transferred to the second funnel, while a new aliquot of mobile phase containing no mixture is added to the first funnel. Both are shaken and equilibrated and the process, called the 'cycle', is repeated until the first aliquot of mobile phase is transferred from the last funnel to a collecting vessel. The procedure can be continued so that a large number of units of mobile phase are collected. The method is simply a scaled-up version of the process shown in Figure 5.1.

This process is obviously very time-consuming and the Craig countercurrent apparatus was developed to mechanize and perform a large series of cycles independently. These machines were very bulky and used large amounts of solvents, so they are not valid for use nowadays for cost and safety reasons.

There have been several developments to reduce time and the amount of solvent used. These are discussed in some detail by Hostettmann, Hostettmann and Marston (1986). Two of the more popular techniques are described below in brief, but detailed instructions regarding use are not given as these machines are expensive and found only in laboratories specializing in extractions.

6.2.3 Droplet countercurrent chromatography

This technique is based on the passage of droplets of the mobile phase through the stationary phase over a long distance. The stationary phase is contained in 200 or more glass tubes connected in series by narrow inert plastic tubes. These tubes are held in racks in the instrument called a droplet countercurrent chromatograph (DCCC). The mobile phase is pumped through as a continuous series of droplets which are small enough to rise (or fall) through the stationary phase without touching the sides of the tube. The components of the mixture, dissolved in the mobile phase, are partitioned between the two phases as the bubble of the mobile phase passes along the series of tubes. Eventually the mobile-phase bubbles reach the end of the series of tubes and are collected as fractions. The components of the mixture which have the greatest affinity for the mobile phase (i.e. partition coefficient in favour of it) are contained in the first fractions to be collected, whereas others are eventually eluted. Any compounds having a strong affinity for the stationary phase will remain in the system, but the stationary phase can be displaced by a new solvent, pumped out and collected. A diagrammatic scheme of the DCCC is shown in Figure 6.5.

Partitioning and subsequent equilibrium is attained rapidly only if a high degree of interfacial activity occurs, and this limits the solvent mixtures that are used. Butanol : acetic acid : water mixtures and chloroform : methanol : water mixtures are those most commonly utilized for polar compounds, but systems have been developed for non-polar compounds. When the solvent combinations specified are mixed and equilibrated, two immiscible phases are formed. One of these is used as the mobile phase and the other as the stationary phase.

The choice of solvent composition to be used for separating a particular mixture is made by using the less polar phase of the solvent combination as the mobile phase in TLC examination of the sample. The sample should resolve into discrete zones. If these are mainly in the upper half of the TLC plate, i.e. R_f values greater than 0.5, then the less

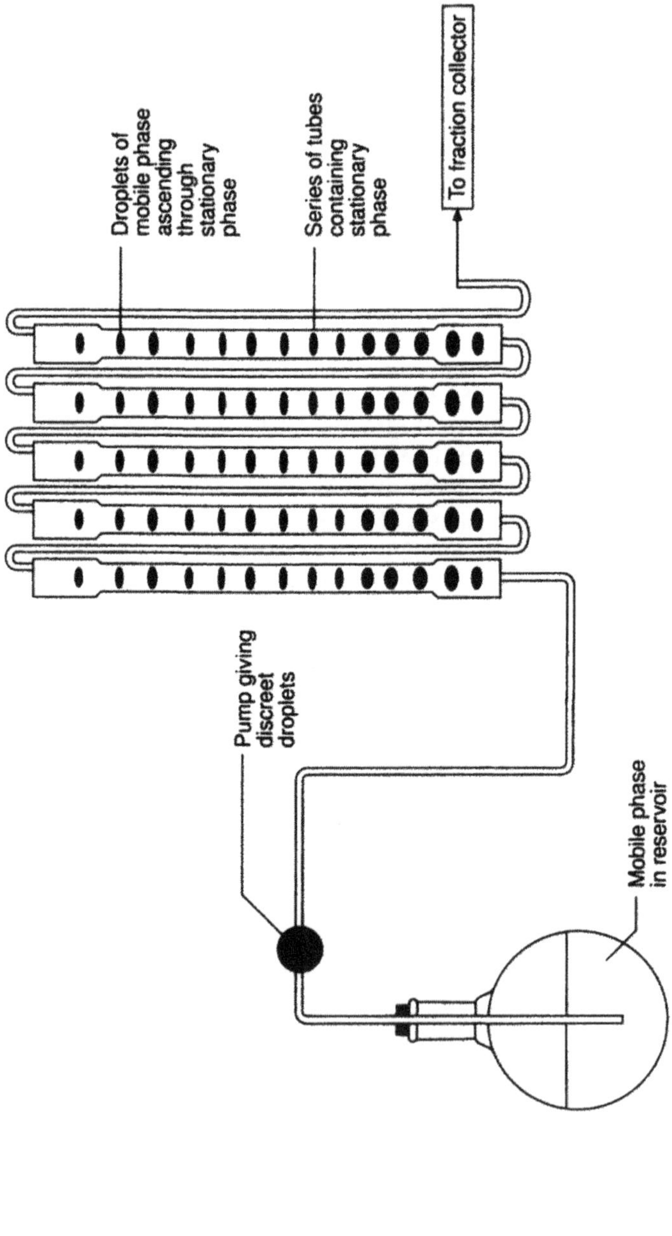

Figure 6.5 Droplet countercurrent chromatography apparatus.

polar phase should be used as the mobile phase in the DCCC. If the components separate with R_f values less than 0.5, then the more polar phase should be used as the mobile phase. The DCCC is designed so that the droplets can either rise (ascending technique) or fall (descending technique) through the stationary phase, and thus either phase can be used as the mobile phase.

DCCC uses relatively low volumes of solvent but it can take several days to separate a mixture. The other major disadvantage of this technique is the limited range of solvent mixtures that can be used.

6.2.4 Centrifugal countercurrent chromatography

Centrifugal countercurrent chromatography is a more recently introduced technique, which yields much faster results than DCCC and is also less restricted in the solvents that can be used.

It consists of a very long tube with many coils. The more-dense stationary phase is held in the tube as a thin layer by centrifugal force as the coils are spun around a central axis. The mobile phase is pumped through the tubes, and so there is a large interface between the two phases across which solute molecules can pass and equilibrium occurs. The mobile phase is collected as time-based fractions as it passes out of the end of the tube.

Using this method, separation of mixtures takes only a few hours. The major limitation of this technique is the cost involved.

6.3 BIBLIOGRAPHY

Braithwaite, A. and Smith, F.J. (1985) *Chromatographic Methods*, 4th edn, Chapman & Hall, London.

Dey, P.M. and Harborne, J.B. (series eds) *Methods in Plant Biochemistry*, Academic Press, London.

Fifield, F.W. and Kealey, D. (1990) *Principles and Practice of Analytical Chemistry*, 3rd edn, Blackie, Glasgow.

Hostettmann, K., Hostettmann, M. and Marston, A. (1986) *Preparative Chromatography Techniques*, Springer, Berlin.

Ikan, R. (1991) *Natural Products – A Laboratory Guide*, 2nd edn, Academic Press, London.

Stahl, E. (1969) *Thin Layer Chromatography*, 2nd edn, George Allen and Unwin, London; Springer, Berlin.

CHAPTER SEVEN

Layer chromatography

Layer chromatography is useful in fractionation, not only as a final process for purification of comparatively small amounts of almost pure compounds, but also as a method for designing some types of column separation and also for monitoring the composition of fractions obtained by other fractionation processes.

It is also a method that is used very widely for qualitative analysis (section 8.2.2). Practical details of basic techniques in thin-layer chromatography are given in section 9.4.

7.1 LAYER CHROMATOGRAPHY

The selection of a suitable system for thin-layer (or paper) chromatography separations follows a logical process, but it should be emphasized that previous experience plays some role in achieving good resolution of the components and that good chromatography is a mixture of art and science! The system selection entails choosing the mobile phase, the stationary phase and the detection method.

Systematic approaches to the choice of mobile and stationary phases have been described in the literature, for example that by Stahl (1969) and the PRISMA method developed by Nyiredy *et al.* (1985), and are discussed below.

7.1.1 Choice of phases

Adsorption chromatography systems (Table 5.2)

Any chromatographic system consists of two phases. In adsorption chromatography the range of stationary phases is limited compared with the large number of mobile phases which can be employed, so only one stationary phase is used in the first instance. (This is almost always silica gel for adsorption chromatography.) Other stationary phases are only tried after attempts with a variety of mobile phases with silica gel have failed to give good results.

The thin-layer plates may be available commercially as ready-made products. If this is not the case, or if cost prohibits the buying of such products, a wide range of stationary phases is available as powders which can be used to generate 'home-made' plates. Details are given in section 9.4.1.

A wide variety of mobile phases is available due to the possibility of mixing two or more liquids to modify the polarity and pH of the mobile phase. Initial studies to achieve resolution of a mixture therefore consist of the application of the mixture to a series of small plates having the same stationary phase. (NB It is better to apply the mixture as a short (4 mm), narrow, horizontal line rather than as a circular spot since this enables the eye to more easily distinguish substances that run very close together on the chromatogram.)

Each plate is then developed in one of a series of mobile phases, usually consisting of single solvents and covering a wide range of polarities. The separation of the components of the mixture is then determined for each system and the two systems giving the best separation noted. A second similar set of separations is then carried out using different mixtures of the two solvents chosen to cover the range of polarities between them. A typical scheme is illustrated in Figure 7.1.

If satisfactory separation is not achieved by this stage then it is a good idea to repeat the process using another adsorbent such as alumina. If preliminary tests have indicated the presence of phenolic compounds, then polyamide may be a better choice as this is known to give good separations for this type of compound.

Adequate separation is often achieved at the end of the second stage, but sometimes further modifications have to be made. The observed zones may not be very discrete and tailing is often observed. Comparatively polar molecules may give zones with this appearance and it is also observed in basic molecules such as alkaloids and acidic compounds such as phenols where the molecule exists in a dynamic equilibrium between its charged and uncharged form. These two forms have different adsorption coefficients and this results in tailing.

In the first instance tailing may be reduced or completely removed by adding a little water to the mobile phase.

In the second instance considerable improvement is often achieved by shifting the pH of the mobile phase to favour much more the existence of the uncharged form. In the case of basic compounds, not more than 5% v/v ammonia or diethylamine is added to the stationary phase, whereas, for acidic compounds, formic or glacial acetic acid is used in the same way.

The PRISMA system introduced by Nyiredy in 1985 combines these features in the range of solvents chosen for the initial investigation of the separation of the mixture. Nine groups of solvents are considered, each having a 'solvent strength' (related to its polarity) and the mixture is run in one example from each group. The solvents in each group are listed in

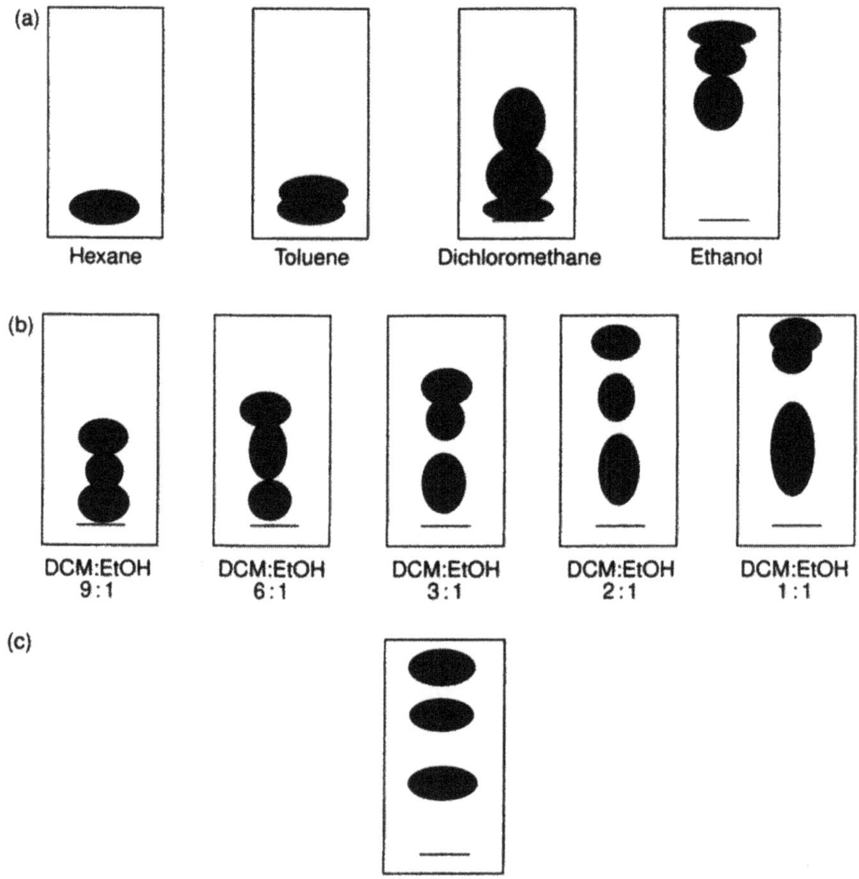

Figure 7.1 Selection of a TLC system. (a) Some separation is seen with dichloromethane (DCM) and ethanol (EtOH). A range of mixtures of these two solvents is then made up and the plates run. (b) The DCM : EtOH 2 : 1 gives the best separation but the zones are elongated. A little water is therefore added to make a mobile phase DCM : EtOH : H_2O 65 : 35 : 5 and the chromatogram obtained. (c) The zones are less elongated and better resolution is achieved.

Table 7.1. The three solvents giving best resolution of the components of the mixture are then drawn as the three vertical sides of a prism, the height of each side corresponding to the solvent strength (Figure 7.2).

The solvent strengths of the two highest sides are then reduced by proportionate adding of hexane to bring them both to the same solvent strength as the lowest of the three. A 1 : 1 : 1 combination of the three adjusted solvents is then used as the mobile phase in a TLC test of the

Table 7.1 Solvent groups for PRISMA TLC optimization

Group	*Solvents*	*Solvent strength*
1	Hexane	0
2	*n*-Butylether	2.1
	i-Propylether	2.4
	Methyl-*t*-butylether	2.7
	Diethylether	2.8
3	*n* Butanol	3.9
	i-Propanol	3.9
	n-Propanol	4.0
	Ethanol	4.3
	Methanol	5.1
4	Tetrahydrofuran	4.0
	Pyridine	5.3
	Methoxyethanol	5.5
	Dimethylformamide	6.4
5	Acetic acid	6.0
	Formamide	9.6
6	Dichloromethane	3.1
	Dichloroethane	3.5
7	Ethyl acetate	4.4
	Butanone	4.7
	Dioxan	4.8
	Acetone	5.1
	Acetonitrile	5.8
8	Toluene	2.4
	Benzene	2.7
	Nitrobenzene	4.4
9	Chloroform	4.1
	Nitromethane	6.0
	Water	10.2

mixture. If the zones run too high, hexane is added until satisfactory R_f values are obtained.

The three component mixtures, including the addition of hexane if appropriate, are then depicted as the three corners of an equilateral triangle, with each side of nine divisions (Figure 7.3).

The mixture tested would correspond to a point in the middle of the equilateral triangle, coordinates being 333. If better resolution is still required, different mixtures are then made, by combining the three components in different proportions which correspond to values nearer the edge of the triangle (e.g. 622), and tested in the TLC system with the sample mixture.

If adequate resolution cannot be achieved, then the whole process has to be repeated using different solvents from groups 1–8.

A hypothetical application of the PRISMA system is displayed in Figure 7.4.

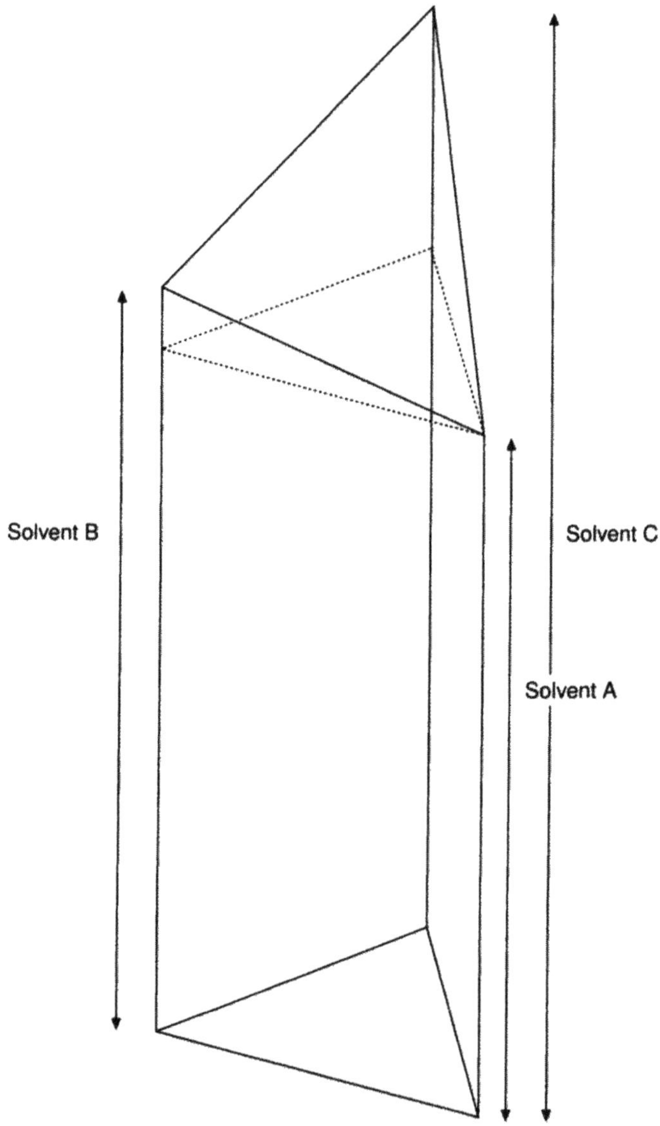

Figure 7.2 PRISMA construction from three mobile phases giving best separation. Height of each vertical side corresponds to calculated solvent strength for each solvent.

Partition systems

The most common layer partition systems used are those where the stationary phase is water, held within cellulose fibres as paper or a thin layer, or reversed-phase silica derivatives.

The approach follows a logical process of testing a range of mobile

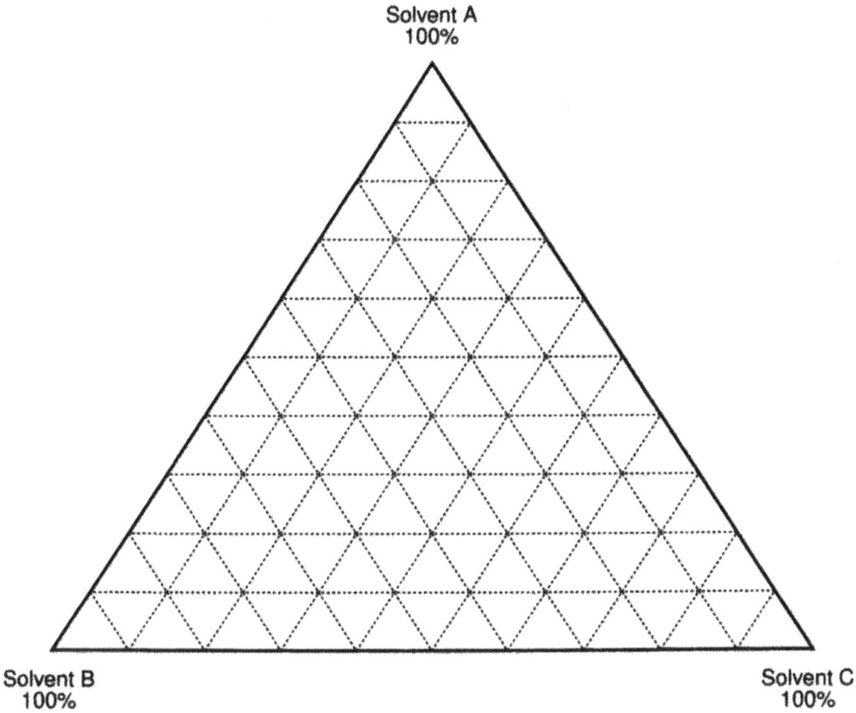

Figure 7.3 PRISMA triangle.

phases with one stationary phase, similar to that discussed above for adsorption chromatography. The solvents are usually much more polar and for cellulose layers should consist of mixtures of water with other polar solvents such as butan-1-ol and glacial acetic acid. For reversed-phase silica systems, the range should extend from water (or buffer solution) alone to methanol or acetone, and perhaps even some dichloromethane : methanol : water mixtures.

The pH of the mobile phase can be varied by the use of buffer solutions or the addition of acid or base to improve resolution.

7.1.2 Choice of detection system

The detection system used in layer chromatography is determined to a large extent by the nature of the substances present in the mixture. These may already be known as the result of preliminary chemical tests (section 8.2.1) but this is not really necessary if a range of detection measures is used.

Although some components of an extract may be coloured, and thus easy to visualize if the stationary phase is white, the vast majority will have little or no colour and other methods have to be used to make them

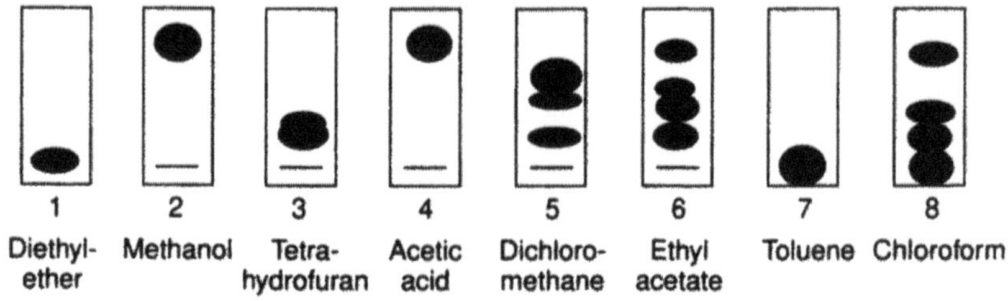

Solvents 5, 6, 8 are the best. Their respective solvent strengths are:

Dichloromethane	3.1
Ethyl acetate	4.4
Chloroform	4.1

The prism model corresponding to this is as follows:

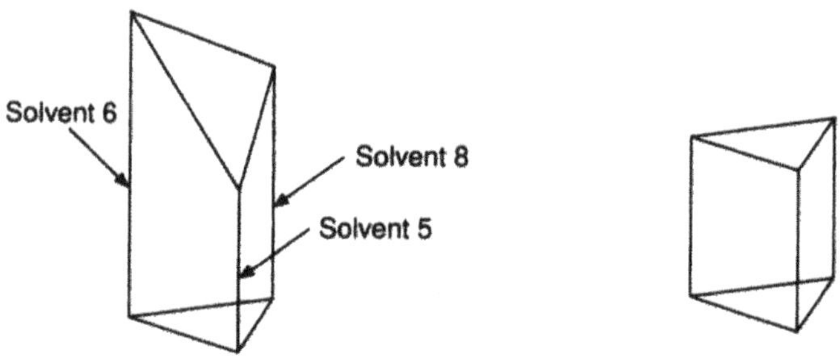

Before adjustment with hexane

After adjustment with hexane

Therefore ethyl acetate and chloroform must be adjusted to 3.1 by addition of hexane.

Chloroform needs addition to make 19.6% hexane, i.e. chloroform : hexane 80.4 : 19.6.
Ethyl acetate needs addition to make 25% hexane, i.e. ethyl acetate : hexane 75 : 25.

A solvent mixture is made as follows:

A	Dichloromethane		1
B	Chloroform : hexane	80.4 : 19.6	1
C	Ethyl acetate : hexane	75 : 25	1

Figure 7.4 (and opposite) Example of the application of the PRISMA method.

visible. The two most common methods are examination under UV light and the use of a spray reagent to produce fluorescent or, more commonly, coloured derivatives. This latter method may also include subsequent heating of the plate.

The result of using this as a TLC mobile phase is:

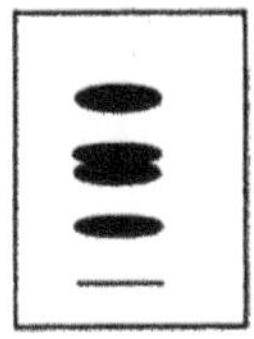

This is not a very good resolution for all the components so some modification is needed.

Construction of a triangle:

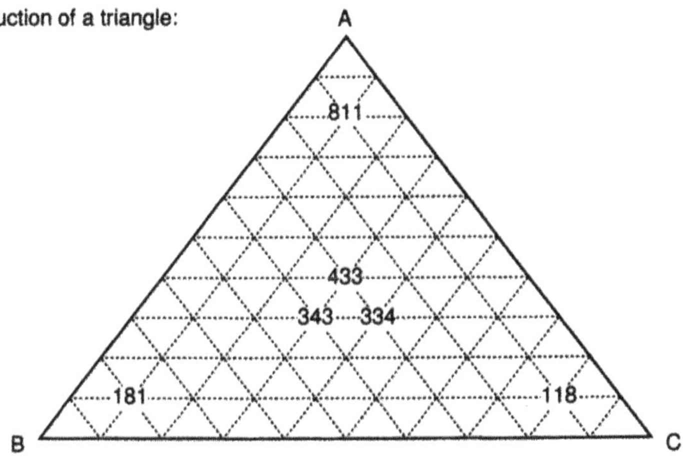

A mobile phase is made up of the following composition:

A	8
B	1
C	1

The resulting chromatogram:

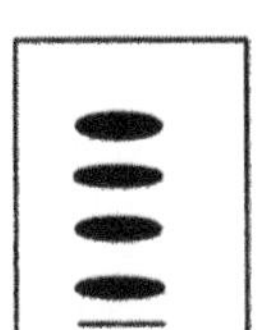

The resolution obtained is very good so that this solvent mixture is used.

Use of UV light

A large number of natural products are either fluorescent (i.e. when exposed to UV light they emit visible light) or absorb UV light. In the latter situation they can be observed as dark areas when examined by UV light if they are coated on a fluorescent layer. The compound will absorb some of the UV light before it reaches the fluorescent substance present in the layer and so the fluorescence is reduced. This effect is known as quenching.

Stationary phases are available which incorporate bonded fluorescent agents and so give a uniform, brightly coloured background when exposed to UV light. These are very useful for the examination of TLC separations and are denoted by the code F_{254} or F_{365}, where the numbers represent the wavelength in nanometres at which the strongest fluorescence occurs. Layers fluorescing at 254 nm are the more frequently used ones.

Ideally, both fluorescent and non-fluorescent layers should be tested in case the extract contains both quenching substances and fluorescent substances which would be masked by the background fluorescence of the layer. It should be noted that some papers have fluorescent substances added during production to enhance their 'whiteness' and substances with weak fluorescence may be difficult to observe if this is the case.

Use of chromogenic or fluorogenic spray reagents

The 'classical' wet chemical tests have been modified to produce a large number of spray reagents which, when applied to a layer, produce colours from colourless substances. Some of these spray reagents are very specific, but many others will react with a broadly based type of compound. A selection of common spray reagents is shown in Table 7.2.

When the best system for separation of a mixture is being investigated and the type of constituents are not known, it is a good example to use several chromogenic reagents, including those containing sulphuric acid which oxidizes many compounds to give coloured derivatives on heating.

7.2 PREPARATIVE THIN-LAYER CHROMATOGRAPHY AND PAPER CHROMATOGRAPHY

7.2.1 Introduction

Fractionation by preparative layer chromatography is limited by the size of the sample that can be applied, usually less than 40 mg total. It is thus most commonly used in the final stages of the purification of pure substances prior to structure elucidation and/or biological testing. A thicker

Table 7.2 Detection systems for layer chromatograms. CARE!!! Always spray chromatograms in a fume hood

Detection method/spray reagent	*Treatment*	*Types of compounds detected*
Methods other than chromogenic sprays		
Daylight		Coloured substances
UV light 254 nm	Use layer which Fluoresces at 254 nm. Examine in darkened area **CARE** Use UV-opaque eye protection	Compounds containing conjugated double bonds
	Use non-fluorescent layer. Examine in darkened area **CARE** Use UV-opaque eye protection	Some fluorescent compounds – usually have extended π electron system
UV light 365 nm	Use non-fluorescent layer. Examine in darkened area **CARE** Use UV-opaque eye protection	Some fluorescent compounds
Exposure to iodine vapour	Use a sealed iodine chamber (iodine crystals in bottom of TLC tank). Leave layer in contact with vapour, then remove. Note yellow/brown zones. These will quickly disappear but can be revisualized by replacing the layer in the chamber OR fixed by spraying with 0.5% w/v starch dextrin solution **CARE** Use iodine chamber in fume cupboard. Do not inhale iodine vapour	Many types of compound, particularly if double bonds are present
Water	Fine spray to wet surface	Lipophilic compounds appear as light zones on a dark background in reflected light, dark zones on light background if held against light

Continued overleaf

Table 7.2 Continued

Detection method/spray reagent	*Treatment*	*Types of compounds detected*
Chromogenic sprays (NB this is a selection only, see bibliography for references giving more comprehensive listings)		
50% Sulphuric acid	Heat at 105°C for 10 min. Observe in daylight	Most organic compounds give a yellow, brown or black zone
5% potassium dichromate in 40% sulphuric acid	Heat at 105°C for 10 min. Observe in daylight	Most organic compounds give a yellow, brown or black zone
0.5% Anisaldehyde in sulphuric acid : glacial acetic acid : methanol 5 : 10 : 85	Heat at 105°C for 10 min. Observe in daylight **CARE** Irritating fumes given off. Use a fume cupboard for heating plate	Terpenoids (usually purple, blue or red) and some other compounds, e.g. some lignans, sugars and flavonoids
10% Antimony trichloride in chloroform	Examine in daylight then under UV light 365 nm. Heat at 105°C for 10 min. Observe in daylight **CARE** Use UV-opaque eye protection when using UV light	Many terpenoids give coloured zones. Some flavonoids fluoresce at 365 nm
0.5% Ninhydrin in acetone	Heat at 105°C for 10 min. Observe in daylight **CARE** Skin turns purple if in contact with ninhydrin. Fingerprints on plate or paper will also be visualized!	Amino acids, peptides give purple/red colours, some secondary amines give yellow colours
5% Ethanolic NaOH or KOH	Examine in daylight	Some phenolic compounds, especially anthraquinones, give red or pink colours
1% Diphenylboric acid 2-aminoethyl ester in methanol, followed by 5% polyethylene glycol 4000 in 96% ethanol	Examine under UV light 365 nm	Flavonoids give fluorescent zones
5% Aqueous iron (III) chloride	Examine in daylight	Phenolic compounds give green or blue colours

20% Aqueous sodium carbonate followed by Folin–Ciocalteu reagent[a]	Examine in daylight	Phenols give blue or purple zones
Dragendorff's reagent[b]	Examine in daylight. Overspray with 5% aqueous iron (III) chloride, 5% aqueous sodium nitrite or 10% sulphuric acid to make zones appear darker	Alkaloids give orange zones on yellow background
0.5% Aqueous Fast Blue B salt followed by 0.1 M aqueous NaOH	Examine in daylight	Phenols and some amines
1% Cerium (IV) sulphate in 10% sulphuric acid	Examine in daylight	Many alkaloids, some other compounds
5% Iron (III) chloride in 335 perchloric acid	Heat at 105°C for 10 min. Observe in daylight **CARE** Strong oxidizing agent. Very corrosive. Irritating fumes given off on heating. Use a fume cupboard for heating plate.	Indole alkaloids give green zones – other alkaloids give various colours

[a] This reagent is available commercially or can be prepared as follows. Add 10 g sodium tungstate and 2.5 g sodium molybdate in a mixture of 70 ml water, 5 ml 85% phosphoric acid and 10 ml conc. hydrochloric acid. Reflux for 10 h. Cool and add 15 g lithium sulphate, 5 ml water and 1 drop bromine. Reflux for a further 15 min, cool and make up to 100 ml with water. **This is the stock solution.** Dilute 10 ml to 40 ml with water before spraying.

[b] This reagent is available commercially or can be prepared as follows. Dissolve 1.7 g basic bismuth nitrate and 20 g tartaric acid in 80 ml water. Dissolve 16 g potassium iodide in 40 ml water. Make a 1 : 1 mixture of the two solutions. **This is the stock solution.** Add 5 ml of the mixture to 50 ml of a 20% w/v solution of tartaric acid in water for the final spraying solution.

stationary phase than that used in analytical techniques is usual for preparative work. Specialized papers are available for preparative paper chromatography (PC). This technique is used primarily for water-soluble compounds and some phenolic substances. Unfortunately resolution tends to be poor so it is not very often used nowadays. The technique for application, separation and removal of fractions from a sample is identical to that described for preparative TLC below.

7.2.2 Conventional preparative TLC

Introduction

Preparative (commonly abbreviated to 'prep') TLC is one of the simplest and cheapest methods available for the isolation of a component or components from a mixture, although it is labour intensive and only small amounts can be obtained from each fractionation procedure.

An example of an exercise involving preparative TLC is given in 10.2.8.

General methodology

The method (Figure 7.5) involves applying the plant extract in the form of a band on a TLC plate, alongside the appropriate reference compounds. The plates used in prep TLC are usually 0.5–1 mm thick (analytical TLC uses plates of 0.25 mm thickness). This allows a greater amount of sample to be loaded on to the plate. Care must be taken to avoid the sample being applied in a diffuse band, as this will decrease the resolution of the components. The plate is developed in a solvent system known to separate the components.

Detection of bands formed

It is extremely important to note that a non-destructive detection method should be used, if at all possible, for the bands separated by preparative layer chromatography. Since detection of analytical TLC is often achieved by destructive chromogenic processes, other methods must be employed, at least in the first instance.

The most common method of visualization is the use of fluorescent layers and examination under UV light, since this would detect all quenching compounds. Similarly, fluorescent compounds can be detected under UV light if an ordinary layer is used.

There are, however, a large minority of naturally occurring compounds that do not fluoresce or quench, so other detection methods have to be used. A simple method for non-water-soluble compounds is to spray the layer with a fine mist of water so that it is wetted enough to become transparent. Non-water-soluble compounds will appear as darker areas

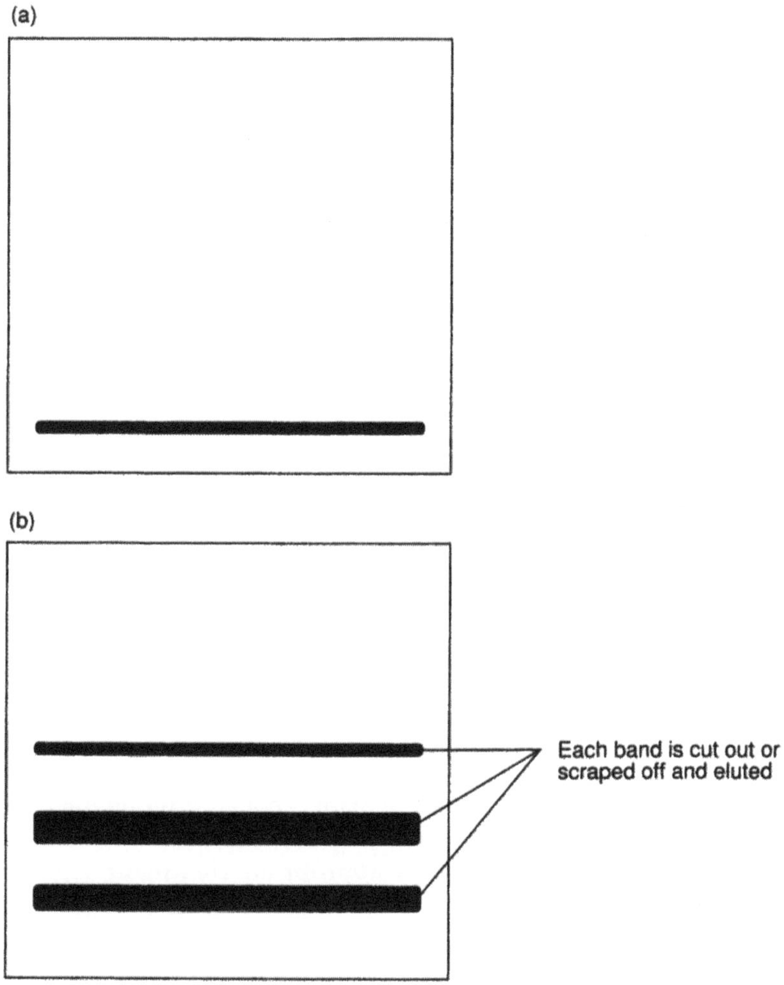

Figure 7.5 Preparative TLC. (a) Before development; (b) after development and visualization.

with transmitted light and as lighter areas when the plate is viewed in reflected light.

Another method that is useful for such compounds is the use of an iodine chamber. This consists of a closed tank into which some iodine crystals are introduced. The atmosphere in the tank becomes saturated with iodine vapour and, when the plate is introduced, iodine preferentially adsorbs to areas of compounds with double bonds in their structure, so that they appear as yellow or brown bands. The colouration is not due to a chemical reaction and, after the zones have been marked, the iodine can be removed by placing the plate in a fume cupboard for

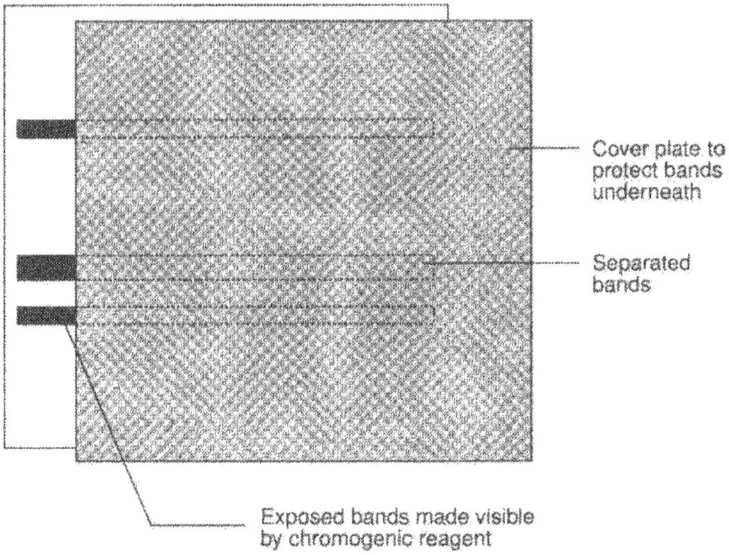

Figure 7.6 Detection of bands on a preparative TLC plate by spraying the exposed edge.

several minutes. (CARE!!! Iodine vapour is irritant to the eyes, nose and mouth.)

If zones cannot be detected by any of these methods, then it may be necessary to use a chromogenic spray but only along a narrow strip at the edge of the plate. At least 90% of the plate is covered while the exposed strip is sprayed (as shown in Figure 7.6.)

The coloured bands appear after appropriate treatment, e.g. careful heating of the exposed strip using a hot air stream. The position of these bands is then used to estimate the position of the bands that had been protected from the spray. These bands only, and not any of the sprayed portion, are marked.

Elution of the band

The marked band is carefully scraped off the plate or cut from the paper or foil and mixed with an excess volume of a suitable solvent. A fairly polar solvent, e.g. ethanol, is usually used to ensure complete elution from the polar silica, but it should be noted that ethanol, methanol and water can dissolve appreciable amounts of silica gel which appear as a residue when the eluant is concentrated. It is best to use the least polar solvent which will completely elute the substance in question. This can be tested by running the mixture on a small TLC plate in the proposed eluant. If the substance moves with the solvent front, then the eluant is suitable for use.

The stationary phase suspension is filtered or centrifuged and the filtrate or supernatant, containing the substance from the band in solution, is evaporated to yield the required compound. A more detailed description of the procedure is given in section 10.2.8.

It is possible to isolate up to about 40 mg of substance using this method. The identity and purity of the compound may then be assessed using spectroscopic techniques. Further purification steps, e.g. recrystallization, may be carried out if necessary.

Conventional preparative thin-layer chromatography has several disadvantages, viz. only about 50 mg of a mixture can be separated on each 20 x 20 cm, 1 mm-thick plate, removal of each band and subsequent elution is time consuming and the silica dust produced is a health hazard. In addition, some of the desired compounds will be lost if detection involves spraying an edge of the plate.

The most successful attempt at reducing or overcoming these problems has been the development of centrifugal layer chromatography, but other techniques have been introduced, e.g. overpressured layer chromatography. Both of these techniques require relatively sophisticated and expensive equipment and so are described here only in outline.

7.2.3 Centrifugal thin-layer chromatography (the chromatotron)

In centrifugal TLC the layer consists of a wide ring of silica (or a similar stationary phase) coated on to a circular glass plate. The silica gel employed contains a large amount of calcium sulphate to improve its binding to the glass, since the plate is rotated at 2400 r.p.m. and centrifugal forces tend to spin the layer off the support unless it is firmly bound. The most common commercial instrument is known as the chromatotron.

The plate is mounted on the spindle of an electric motor and enclosed in a gas-tight chamber through which nitrogen is blown at slight positive pressure. The developing solvent is chosen by first finding a suitable mobile phase to give good separation of the sample mixture using conventional TLC. The eluting solvent for the centrifugal separation should be made slightly less polar than the one found best by ordinary TLC.

Once chosen, the developing solvent (and the sample mixture in solution) is introduced via a wick on to the glass plate surface to the inside circular edge of the layer. As the plate is rotating, the solvent is spun outwards by centrifugal force and passes through the layer (Figure 7.7).

The sample is applied in the same way so its 'starting position' for chromatographic separation is the inside edge of the ring of layer. As the developing solvent is fed through the wick and thence through the layer, separation of the components of the mixture occurs by the normal processes operating in TLC, but the rate of movement is much greater. Components of the mixture separate as concentric rings which gradually move towards the outside of the layer. When a ring containing a

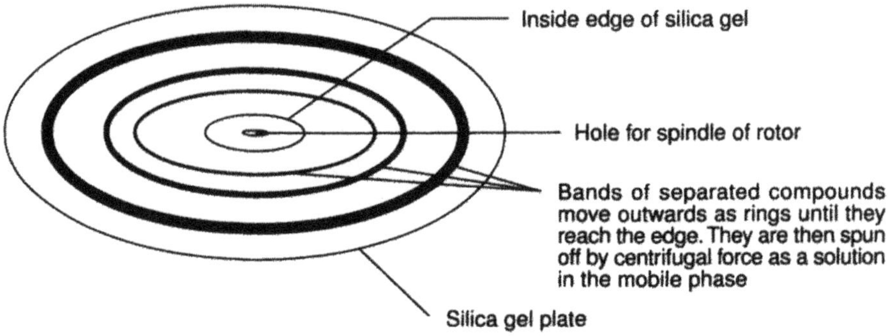

Figure 7.7 Separation of substances by centrifugal chromatography.

component reaches the outside of the layer it is spun off, dissolved in the mobile phase. The inside of the chamber has a gutter and the mobile phase collects in this and drains through a small hole at the bottom. The bands can be collected as fractions by a tube leading from the drain hole. Fractions are analysed by, for example, TLC, and like fractions bulked and left to crystallize or subjected to further purification.

If necessary, gradient elution can be used to speed up the elution of the slower-running components of the mixture.

7.2.4 Overpressured layer chromatography

This technique is, in effect, flat-bed column chromatography, since there is no vapour space above the TLC plate. Diagrams of the apparatus are given in Figures 7.8 and 7.9.

The layer, usually of the same thickness as used for ordinary preparative TLC, is covered with a flexible membrane under positive pressure. This is usually achieved by a bag, made of a substance inert to the developing solvent and compounds to be separated, filled with a suitable fluid so that its outer surface conforms to the irregularities of the layer and thus eliminates any gas space. The mobile phase is pumped through the layer and delivered evenly through a narrow slit to one edge of the plate. The edges of the plate parallel to the flow of the solvent are sealed to prevent leakage of the mobile phase, using a suitable impermeable impregnation agent, so that the mobile phase flows evenly with a horizontal front through the layer. The developing solvent eventually reaches the end of the plate and, because of the pressure from the pump, is collected into an elution tube through a detector. The components of the mixture move as horizontal bands through the layer. Each band in turn reaches the end of the plate, elutes with the mobile phase and can be detected and collected.

As with centrifugal layer chromatography, this technique overcomes

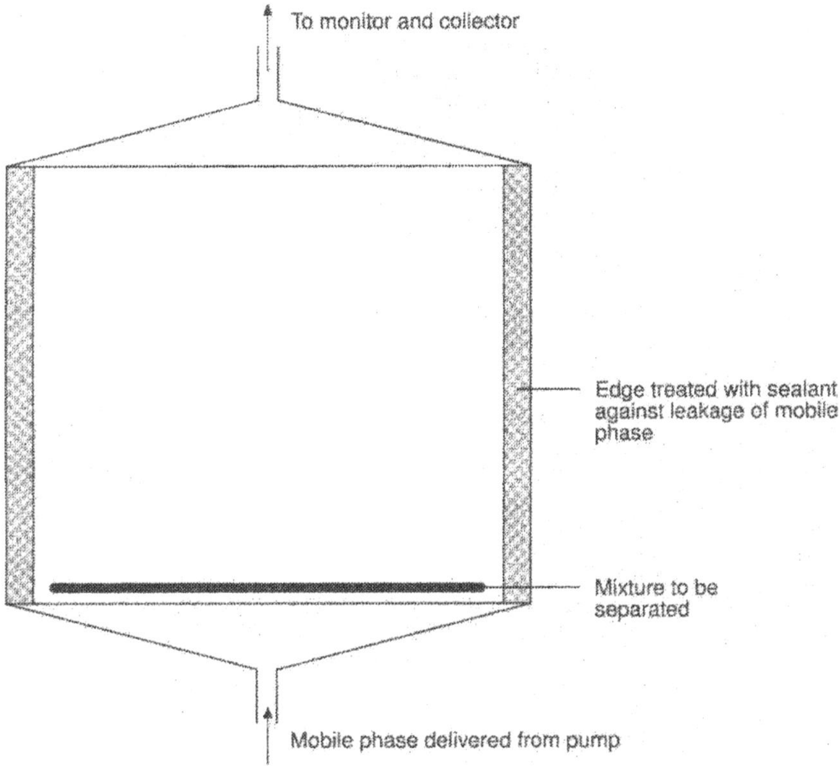

Figure 7.8 Overpressured TLC.

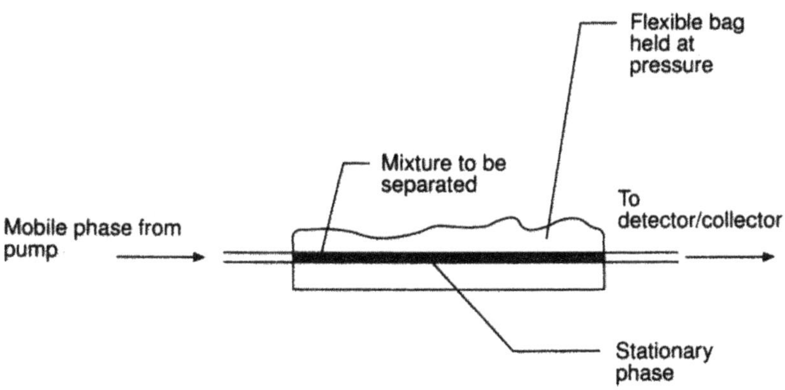

Figure 7.9 Overpressured TLC (side view).

the need to scrape off and elute bands and the plates can be re-used. Its major disadvantages are the relatively small amount of mixture, less than 50 mg, that can be separated at any one time and the sealing of the edges of the plate to completely prevent any side leakage.

7.3 BIBLIOGRAPHY

Braithwaite, A. and Smith, F.J. (1985) *Chromatographic Methods*, 4th edn, Chapman & Hall, London.

Dey, P.M. and Harborne, J.B. (series eds) *Methods in Plant Biochemistry*, Academic Press, London.

Fifield, F.W. and Kealey, D. (1990) *Principles and Practice of Analytical Chemistry*, 3rd edn, Blackie, Glasgow.

Hostettmann, K., Hostettmann, M. and Marston, A. (1986) *Preparative Chromatography Techniques*, Springer, Berlin.

Ikan, R. (1991) *Natural Products – A Laboratory Guide*, 2nd edn, Academic Press, London.

Kinghorn, A.D. and Balandrin, M.F. (1993) *Human Medicinal Agents from Plants*, American Chemical Society Symposium Series 534, pp. 1–348.

Nyiredy, Sz., Erdelmeier, C.A.J., Meier, B. and Sticher, O. (1985) The PRISMA mobile phase optimisation model in thin layer chromatography – separation of natural compounds. *Planta Medica*, **51**, 185–246.

Nyiredy, Sz., Dallenbach-Tolke, K., Erdelmeier, C.A.J. *et al.* (1985) Abstracts 33rd Annual Congress of the Society for Medicinal Plant Research, Regensburg.

Stahl, E. (1969) *Thin Layer Chromatography*, 2nd edn, George Allen and Unwin, London; Springer, Berlin.

CHAPTER EIGHT

Analysis of crude extracts, fractions and isolated compounds

8.1 INTRODUCTION

The preceding chapters have illustrated the different techniques that may be used in order to obtain crude extracts (complex mixtures), fractions (simpler mixtures) and isolated (pure) compounds from natural sources. These are usually then subjected to a number of further analytical investigations in order to obtain more information on the properties of their constituent substances.

Broadly, the nature of further investigations on a sample will be of three types:

- Qualitative chemical analyses – determination of the nature of the constituents of a mixture or the structure of an isolated compound.
- Quantitative chemical analyses – determination of the purity of an isolated substance or the concentration of a single substance or group of substances in a mixture.
- Bioassays – determination of the biological or pharmacological activity of substances and the dose range over which they exert their effects.

The amount of fraction available is possibly the most important factor in making decisions about its future treatment and analysis. Investigative methods can be either non-destructive or destructive. Non-destructive methods mean that the sample can be recovered (perhaps with a slight clean-up procedure, as shown in Table 8.2) and used for other tests. Some physico-chemical procedures, e.g. chemical tests, analytical chromatography, mass spectrometry, and all biological testing procedures are destructive, i.e. they use up the sample and generally it cannot be recovered.

Another important factor is that the different types of analysis require the sample to be present in different types of medium. It is always desirable to supply the fractions as solids which can be weighed accurately and reconstituted in the appropriate solvents. The solid forms are usually produced from solutions by evaporation under reduced pressure to minimize decomposition or, in the case of aqueous solutions, by freeze-drying, sometimes called lyophilization (section 9.1.2).

This present chapter will address the methods involved in some of the types of analysis to which extracts and isolated compounds may be subjected, and the practical considerations associated with their use.

8.2 QUALITATIVE ANALYSIS OF MIXTURES AND ISOLATES

The nature (chemical group, specific identity, polarity, etc.) of the substances in a mixture can be determined by a number of means including 'wet' chemical tests and chromatography. In chemical tests, a colour reaction or precipitate in response to specific reagents may indicate the presence of a particular compound or, more usually, a class of compound. Such tests can be useful to investigate the chemical constituents of test organisms and also to monitor the effectiveness of an extraction process when a particular chemical class is being sought. The behaviour (mobility) of substances in chromatographic systems is usually highly reproducible (if conditions remain constant) and this can yield information on their identity and /or physico-chemical properties, as well as the identity of the source organism. In thin-layer chromatography, chromatographic behaviour can be coupled with chemical tests, since many of the reagents used in wet chemical tests can be used as spray reagents, thereby confirming the chemical nature of a particular zone.

Where the identity of an isolated compound is suspected with a high degree of certainty, it may be confirmed by means of chemical reactions and chromatographic behaviour. For isolated compounds whose structure is less certain, additional methods such as mass spectrometry (MS) and ultraviolet, infrared and nuclear magnetic resonance spectroscopies (UV, IR, NMR) can be applied in order to elucidate unambiguously the molecular structure of the compound. These different forms of qualitative analysis are discussed further in this section.

8.2.1 Chemical tests for different chemical classes

Tests are available to enable the presence of particular chemical groups to be detected in extracts. These are based on a colour reaction or precipitation in response to a particular reagent. An important point to note is that although these tests give an indication of the chemical groups present, they are seldom specific enough to enable unequivocal conclusions to be reached. It will be apparent from Chapter 3 that in order to test any organic material for the presence of compounds of a specific chemical class, an appropriate extraction procedure has to be adopted. Extracts may also need to be concentrated down in order for a visible reaction to take place. A detailed guide to the test procedures is given in section 10.1.

Many of the chemical tests given here are applied as identification tests in pharmacopoeial monographs for crude drugs known to contain those constituents, and can be adapted for use as TLC spraying agents or to produce coloured solutions for quantitative analysis.

Fixed and volatile oils

Volatile oils consist mainly of a mixture of small molecules of hydrocarbons containing 10 or 15 carbon atoms, whereas the molecules found in fixed oils are much larger and consist of esters of glycerol and long-chain aliphatic acids. Their solubility properties are similar and thus both types of oil are extracted by non-polar solvents such as petroleum ether or dichloromethane. When applied to a piece of filter paper, both oils will leave a translucent grease spot (after evaporation of the extracting solvent). The two types of oil can be distinguished in that volatile oils have a strong smell, whereas fixed oils do not. Additionally, when the filter paper is heated in an oven, a grease spot due to a volatile oil will disappear whereas that of a fixed oil will remain.

Carotenoids

The presence of carotenoids in extracts may be indicated by a bright yellow, orange or red colour. Many carotenoids give a deep blue colour when treated (in chloroformic solution) with concentrated sulphuric acid or with 5% antimony trichloride in chloroform.

Alkaloids

Large numbers of alkaloids have been isolated from natural sources and many different structural subgroups exist which vary in the type of ring skeleton in which the nitrogen is incorporated. A general test relies on the fact that most of these alkaloids, in acid solution, form precipitates with heavy-metal reagents such as Dragendorff's reagent (potassium iodobismuthate; orange precipitate) and Mayer's reagent (potassium iodomercurate; white or buff precipitate). Other reagents have been found which give colours with particular alkaloidal subgroups or particular alkaloids. The Vitali–Morin test is not specific for tropane alkaloids, e.g. hyoscyamine (scopolamine, Figure 8.1) but may be used in their detection. It involves nitration of the benzene ring found in many such alkaloids, and colours are formed on subsequent addition of alkali.

Ehrlich's reagent (4-dimethylaminobenzaldehyde solution) is a strong electrophile which reacts with the electron-rich α-carbon of indole rings to form a blue-coloured adduct. It can be used to detect the presence of certain indole alkaloids, e.g. ergotamine (Figure 8.1) where steric or

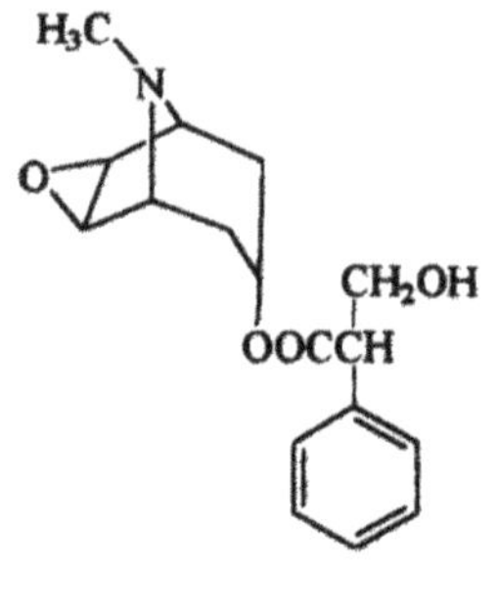

Scopolamine, a tropane alkaloid

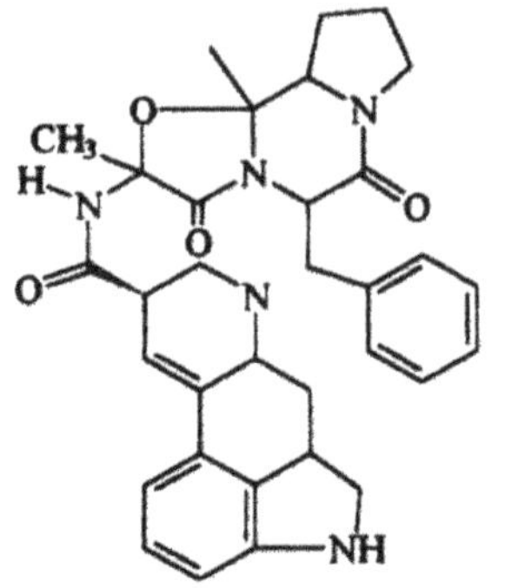

Ergotamine,
an indole alkaloid

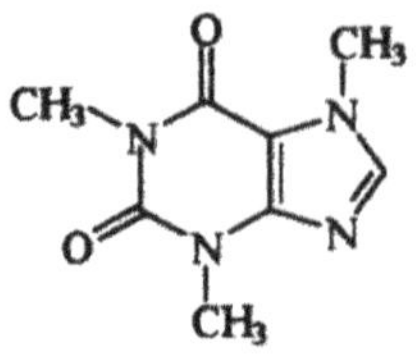

Caffeine,
a xanthine

Figure 8.1 Examples of alkaloids.

chemical factors allow the reaction to occur. Other indole alkaloids, e.g. strychnine and brucine, do not give this colour reaction.

The xanthine alkaloids (e.g. caffeine, Figure 8.1) do not respond to the standard test with Dragendorff's or Mayer's reagent. However they respond to the murexide test, which involves reaction with an oxidizing agent, e.g. hydrogen peroxide or potassium chlorate (red colour formed), followed by exposure to ammonia solution or vapour (colour changes to red-violet).

Glycosides

Glycosides also belong to a wide range of chemical subgroups. The majority of these share common properties,, such as solubility in water or aqueous alcohol and susceptibility to acid hydrolysis to yield the aglycone.

The general test for the presence of glycosides involves parallel testing of a water extract and an acid extract of the organic material. Water extracts intact glycosides and any free reducing sugars already present in the material. Treatment with acid results in the hydrolysis of *O*-, *N*- and *S*-glycosides to one or more reducing sugars and the free aglycone. The newly released sugar, along with any free sugars pre-existing in the plant material, will be extracted into the acid. Both extracts are tested with Fehling's reagent, which contains copper ions. In the presence of reducing sugars, Cu^{2+} ions are reduced to Cu^{+} and a red precipitate of copper oxide (Cu_2O) is produced. A heavier precipitate will be produced with the acid extract if glycosides are present in the sample material. However, this method does not apply to C-glycosides.

A number of semispecific tests have been devised which rely mostly on the special properties conferred by the aglycone moiety. Anthracene derivative glycosides (e.g. Frangulin A, Figure 8.2) may be detected by

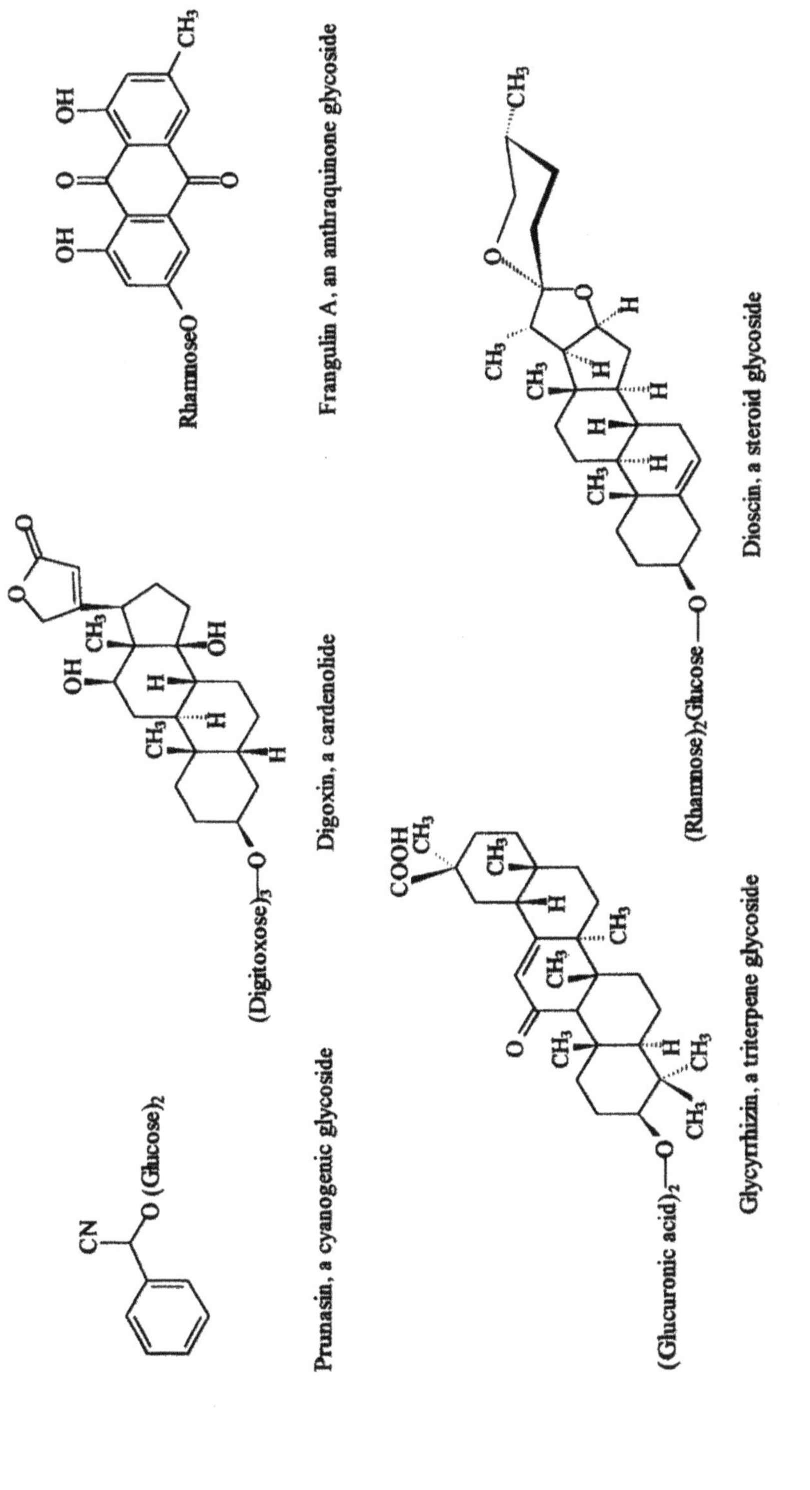

Figure 8.2 Examples of glycosides.

the pink colour produced by dihydroxyanthraquinone aglycones in alkaline solution. Acidic ferric chloride may be used to oxidize reduced anthracene forms to the anthraquinone structure. This reagent also hydrolyses C-glycosides which are common amongst this group and are resistant to hydrolysis by acid alone.

Cardenolides are a group of cardioactive glycosides in which the aglycone consists of a steroid linked to an α,β-unsaturated lactone and the sugar chain generally includes deoxy-sugars (e.g. digoxin, Figure 8.2). These two structural features give colour reactions in the Kedde and Keller–Kiliani tests, respectively (section 10.1.4). Alcohol is often used to extract cardenolide glycosides (which also display solubility in chloroform). This may result in the co-extraction of chlorophyll and phenolic pigments which can be removed by lead subacetate precipitation (section 3.5.2) before the chemical tests are carried out.

Saponin glycosides are of two types – those based on a triterpenoid aglycone and those with a steroidal aglycone (e.g. glycyrrhizin and dioscin, respectively; Figure 8.2). The combination of a bulky, lipophilic aglycone with hydrophilic sugar moieties confers surface active properties on the molecules ('saponin' = soap-like), and consequently their presence in aqueous extracts can be inferred by the formation of stable froths on shaking. A related property is their ability to haemolyse red blood corpuscles (resulting from disruption of the cell membrane). This can also be used as a basis for a test, although it works better with steroidal saponins than triterpenoid derivatives that have a slower haemolysing action. The presence of cyanogenic glycosides (e.g. prunasin, Figure 8.2) can be investigated by a method based on their decomposition to yield hydrogen cyanide. Plant material containing these glycosides commonly contains glycosidase enzymes which are capable of cleaving sugars from the aglycone. Applying a small quantity of water to the plant material can activate these enzymes which then break the glycosides to release hydrogen cyanide. This reacts with sodium picrate to form the reddish purple compound, sodium isopurpurate.

Phenolic compounds

This group includes flavonoids, tannins and other phenols (Figure 8.3). Due to the presence of vicinal oxygenated functions, many of these compounds can act as ligands for metal ions. Thus ferric ions when added to solutions of phenols produce dark green, blue or blue–black complexes. Phenolic compounds can exist in free or glycosidic form.

Terpenoids

Terpenoids give a positive reaction (colours ranging from red to blue) when warmed with acetic anhydride and treated with a few drops of concentrated sulphuric acid (Liebermann-Burchard reaction).

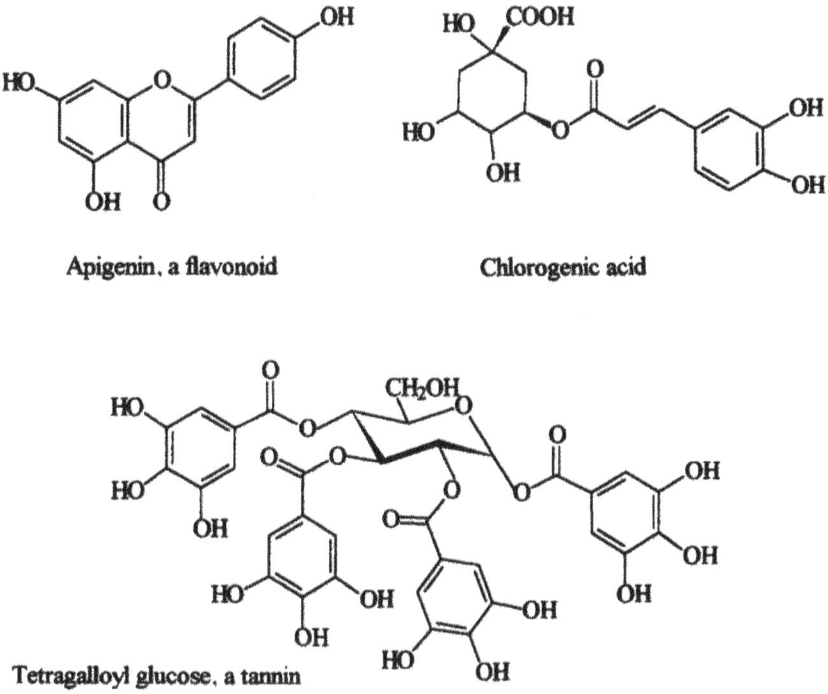

Apigenin, a flavonoid

Chlorogenic acid

Tetragalloyl glucose, a tannin

Figure 8.3 Examples of phenolic compounds from plants.

Sugars, polysaccharides, amino acids and proteins

The Fehling's test can be applied to determine the presence of reducing sugars in extracts, and the presence of sugar polymers may be inferred from the generation of reducing sugars after acid hydrolysis of the extract. Polysaccharides may also impart a gel-like consistency to aqueous solutions and a swelling index is often used to determine their presence. Ninhydrin reagent gives purple, pink or red derivatives with most amino acids, although the proline derivative is yellow. Proteins may be hydrolysed to amino acids for testing with ninhydrin or, alternatively, dyes such as naphthalene black 12B or Coomassie blue, which adsorb on to proteins, may be used for their detection.

8.2.2 Qualitative analysis using chromatographic methods

Thin-layer chromatography

The most useful form of chromatography for qualitative analysis of crude extracts or isolated compounds is thin-layer chromatography (TLC). This technique is used widely to identify isolated substances or constituents present in an extract by comparison with reference substances and/or data in the literature (Figure 8.4). For a given substance, a useful parameter to measure is its retardation factor or the R_f value of its zone on TLC. This is the ratio of the distance from the baseline (point of application) to the centre of the zone divided by the distance from the baseline to the solvent front.

In TLC, the identification of a reference substance as being present in an extract is only valid if the following criteria are met.

1. The substance and reference compound exhibit identical R_f values (section 7.1) in every TLC system tested.
2. Several different detection methods are used (section 7.1.2) and the substance gives identical reactions to the reference substance with all the detection methods used.
3. At least five different mobile phases are used to determine a range of R_f values.

Spray reagents based on those giving the colour reactions described in section 8.2.1 may be useful in determining the nature of the compounds forming zones on the TLC plate. A number of common spray reagents are given in Table 7.2.

When comparing R_f values of a component with those quoted in the literature, or those recorded in one's own previous experiments, it is important to remember that variations may occur due to slight changes in temperature, stationary phase or mobile phase composition and saturation status of the TLC tank. To prevent ambiguity, it is a good practice to run a marker substance, preferably with an R_f value of about 0.5, alongside the samples, and express the R_f values of all test substances relative to that of the marker. Thus a zone running at the same R_f value as the marker has a relative R_f value of 1, those above it have relative R_f values greater than 1 and those below it have relative R_f values less than 1. Relative R_f values are more reliable than absolute figures for comparing chromatograms (R_a, Figure 8.4).

TLC can also be used in the identification of the organism from which an extract has been derived. Some organisms contain a complicated mixture of related components and in these cases the identification of the organism may be helped significantly by comparison of the 'zone profile' (similar in principle to a fingerprint) of an extract with that given by an authentic sample. It is important to extract both sample and reference material by identical procedures. The profiles of the zones should be virtually identical both qualitatively and quantitatively, i.e. the same zones should be present in about the same relative intensities. Comparison of zone profiles may be important when several batches of a particular organism under study are obtained at different times and it is important to establish that they are the same species. However, it should be noted that biological variation due to a variety of factors may cause the zone

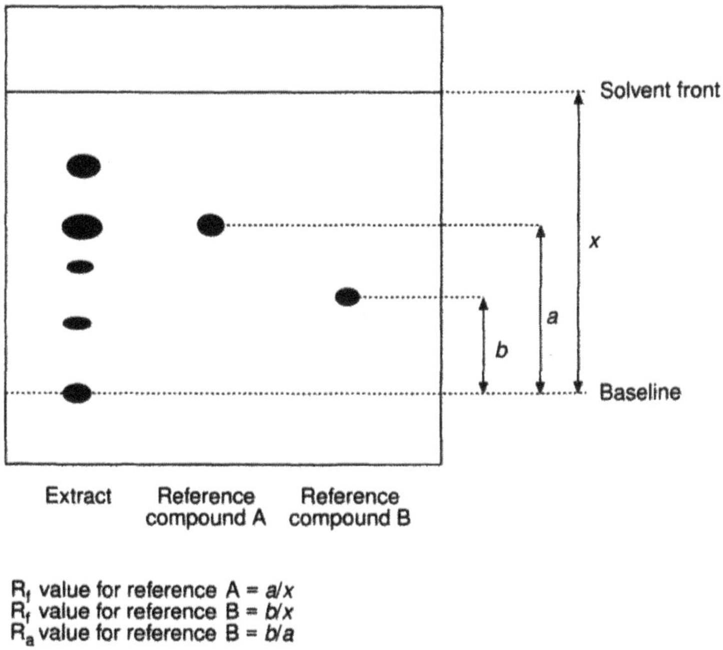

Figure 8.4 Typical TLC chromatogram. From comparison of R values, the extract contains at least five compounds, one of which may be A, whereas B is absent in detectable amounts.

profiles from the same plant species to differ. This method is therefore not to be recommended for identification of an organism or any other substance without supporting evidence from microscopy or other techniques.

TLC may also be useful in revealing the presence of adulterants, such as a closely related species, in a sample. For example, medicinal rhubarb (*Rheum officinale*) may be substituted or adulterated with rhapontic rhubarb (*Rheum rhaponticum*) which contains the oestrogenic stilbene derivative rhaponticin. The presence of an additional zone or band corresponding to this compound in the zone profile indicates that the sample is not purely medicinal rhubarb. Similarly, a zone corresponding to carvone in peppermint indicates adulteration with spearmint.

Gas and liquid chromatography

Other forms of chromatography used in the analysis of natural products include gas chromatography (GC) and high-performance liquid chromatography (HPLC). Although mainly used for quantitative work (section 8.3.4), they can also yield useful qualitative information. In both these techniques, the sample is applied, via an injection port, to the top of

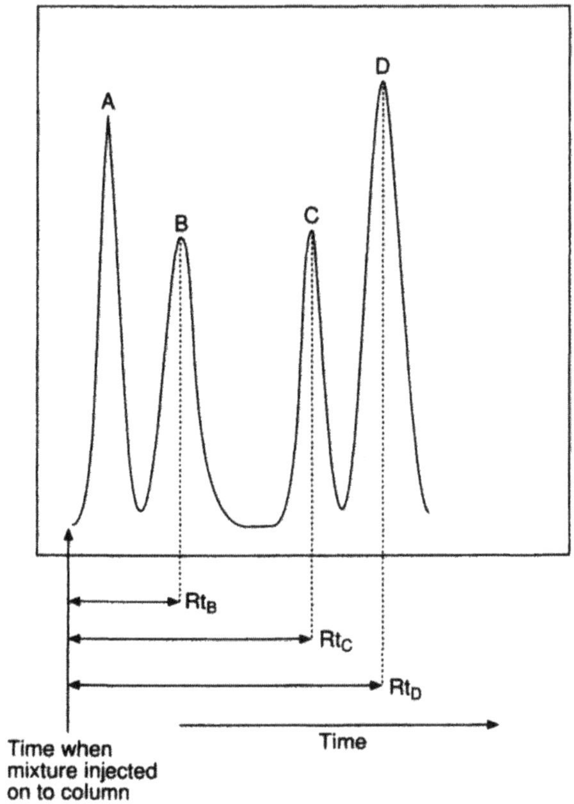

Figure 8.5 Typical HPLC chromatogram. A reference having the same retention time, R_t, as one of the peaks would indicate that the reference compound may be present in the mixture. A, B, C, D are separated compounds.

a column through which a mobile phase is flowing continuously. As with TLC, separation of the components in the sample is achieved due to differences in their relative affinity for the stationary and mobile phases. The component with the greatest affinity for the mobile phase will elute from the column first, whereas that with the greatest affinity for the stationary phase will elute last. Eluting substances are monitored with a suitable detector and the output plotted as a graph of detector signal versus time. This is the GC or HPLC chromatogram (Figure 8.5), and typically consists of a series of peaks corresponding to the number of separated components.

The time between the introduction of a sample and the tip of the peak of a particular component is called the retention time (R_t), of that component. Strictly speaking, the true R_t is the difference between the time taken for a particular substance to emerge and the time taken for the mobile phase to travel from the injection port to the detector. This is a constant characteristic for each substance under given chromatographic conditions. Thus R_t values of components in a sample can be compared with those of reference standards to obtain a tentative identification. As with TLC, substances should show matching R_t values in a series of systems before their identity with a reference substance can be confirmed. Where reference compounds are not available, or where chromatographic conditions tend to vary, R_t values relative to that of a particular component or added standard may be a more reliable parameter for comparison than absolute R_t values. A tentative identification can be further substantiated by injecting a mixture of the unknown sample and the reference compound and examining the chromatogram to determine whether a single, symmetrical peak is produced.

The order of elution of substances from a column can give information on their relative polarity, pK_a, or molecular weight, depending on the nature of the chromatographic system being used. As with TLC zone profiles, GC or HPLC peak profiles can give information on the identity of the starting material. A brief outline of the principles of GC and HPLC will be given here; the reader should consult more specialist textbooks for greater detail of instrumentation and the scientific literature for methods pertaining to specific chemical classes.

In gas chromatography, the stationary phase is usually a liquid, either coated on to the walls of a narrow column (as in capillary GC) or on to small particles of an inert, solid material which is packed into a wider column (packed-column GC). The mobile phase is an inert gas (often nitrogen, helium or argon) and the components of the sample mixture separate due to differences in their relative affinities for the liquid and gaseous phases. The columns have to be held at elevated temperatures, typically 100–400°C, in order to ensure that samples have sufficient tendency to enter the gaseous phase to elute from the column. Thus this technique is most suitable for thermostable, volatile substances such as volatile oils. Compounds with hydroxy groups or carboxylic groups, e.g. sugars, fatty acids, sterols and phenols, generally chromatograph very poorly on direct injection, due to their low volatility and interactions with the silicaceous material forming the inert support. However, it is possible to make derivatives of substances using methylating, acylating and silylating agents, which improves volatility, thermostability and chromatographic characteristics. A range of stationary phases of varying polarity are available to suit the polarity of the compounds to be analysed (Table 8.1).

The most common detection method used – the flame ionization detector (FID) – is a destructive method but is universally applicable. As a result, a very wide range of natural products may be analysed by GC, either directly or as derivatives. In practice, however, HPLC may be preferred for many substances, particularly where thermolability or selective

Table 8.1 Some stationary-phase materials used in gas chromatography (some names used are brand names)

Stationary phase	*Polarity*
Squalane, Squalene	Non-polar
Apiezon L	Non-polar
Silicone oils, e.g.	Variable
SE-52 (phenyl-)	Non-polar
SE-30 (methyl-)	Non-polar
XE-60 (cyano-)	Semi-polar
Carbowax	Polar
Polyethylene glycol esters (e.g. DEGS)	Polar
Polyethylene glycols	Very polar

detection is an issue. GC is relatively easily coupled to a mass spectrometer as detector (GC–MS), giving the added possibility of identifying peaks by both R_t and mass spectrum.

High-performance liquid chromatography uses most of the stationary and mobile phases common to preparative column chromatography (Chapter 6), although the particle size is considerably smaller (2–10 μm), resulting in improved separation of closely related substances. Consequently, the liquid mobile phase needs to be pumped through the column at elevated pressures (typically 100–300 bar) to achieve practicable flow rates. HPLC is usually run at room temperature, though it is possible to cool or heat the column for individual applications. The commonly used stationary phases are similar to those used for preparative chromatography (section 6.1) and include C_{18}-silica (ODS, reverse phase) and silica (normal phase), which are generally suitable for a wide range of phytochemicals. Ion-exchange bonded silica packing materials are also used. Due to the instability of silica columns at pH values outside the range 3–7, more resistant polystyrene resin-based reversed-phase and ion-exchange columns are also available but are more expensive. HPLC gel-filtration columns are based not on dextran gels but on silicaceous beads or polystyrene resins as these are better able to withstand the high pressures in the system. A range of stationary phases has been developed to separate optical isomers without the need to prepare diastereomeric derivatives of the compounds (Dappen, Arm and Meyer, 1986). These systems may be based on cyclodextrins, chirally selective proteins bonded to agarose, or 'brush' type phases consisting of one isomer of a chiral molecule bonded on to silica. The mobile phases used in HPLC range from organic solvents (normal phase chromatography) to mixtures of water or buffer with water-miscible organic modifiers such as methanol, acetonitrile and tetrahydrofuran.

Unlike the FID detector in GC, the most common type of HPLC detector, the ultraviolet–visible (UV–VIS) detector, is not universally applicable. It detects compounds based on their ability to absorb light in the visible or UV range. The column eluant can either be monitored at a single wavelength or, if a stop-flow scanning detector or a diode array detector (LC-DAD) is used, at a number of wavelengths from which the UV–VIS spectrum of eluting compounds can also be recorded. The absorption spectrum gives added qualitative information about the sample. It can also be used to determine whether a peak consists of a single substance or two or more unresolved materials, by comparing the shape of the spectrum at different time points within the peak. Fluorescence detectors give an enhanced level of selectivity and sensitivity.

For compounds without a chromophore or fluorophore, derivatives can be made, e.g. by condensation with chromophoric or fluorometric substances such as dansyl chloride (dansylation). Alternatives such as refractive index (useful for sugars), electrochemical and evaporative light scattering detectors (a universal detector) may be used. As with GC, mass spectrometry may be interfaced with HPLC (LC–MS) but this technique has posed greater difficulties than GC–MS due to the limited range of MS-compatible HPLC mobile phases. The UV–VIS and refractive index detectors are non-destructive methods, which means that the samples eluting from the column may be collected and recovered by removal of the mobile phase.

The advantages of GC and HPLC as compared with TLC are the smaller quantities of sample required for analysis, far superior resolution (separation, Figure 5.3) of substances, the existence of a variety of stationary phases to suit different applications, and the ability to obtain accurate quantitation of substances present in samples (section 8.3.4). Disadvantages include the greater cost of materials and equipment, and the need to analyse samples in series rather than in parallel. Due to the high resolution achieved, HPLC may also be used on a semipreparative scale to perform the final purification of a substance from closely related impurities such as isomers.

Sample preparation for chromatographic analysis is an important consideration. Due to the small volumes needed (typically 1–50 μl for TLC; 1–5 μl for GC and 5–100 μl for HPLC), it is important that the sample solution is sufficiently concentrated for the amounts applied to be detectable. For TLC it is desirable that the sample is dissolved in a volatile solvent so that diffusion into the layer is minimized during application, although water can be used if there is no other option. For GC, the use of water is to be avoided as it lacks volatility and furthermore can interact with the silica of the adsorbent. For HPLC, the mobile phase is to be used as the sample solvent wherever possible, and in all cases the sample solvent and the mobile phase must be miscible.

In attempting to analyse a sample by TLC, GC or HPLC, a lot of effort in method development can be saved by consulting the scientific literature for systems used by others for the same or related chemical compounds. Details to note carefully are the stationary phase (and support material for GC), mobile phase, conditions (e.g. temperature), column or layer dimensions and detection methods. It is often found that, due to variation in the materials and apparatus available in different laboratories, some modification of published methods needs to be undertaken before ideal conditions for separation are achieved.

8.2.3 Qualitative spectroscopic methods and structure elucidation

A number of spectroscopic methods are available for the determination of the chemical structure of an isolated compound. The majority of these physico-chemical analytical methods (e.g. determination of UV, IR and NMR spectra) are non-destructive since they utilize the sample dissolved in an appropriate solvent. As long as the sample is a pure compound, these tests should always be carried out before final biological tests are performed. Methods that can be used to recover the compound from the test matrix are given in Table 8.2. Mass spectrometry is a destructive method, but fortunately requires the least quantity of sample of the various methods described here.

Table 8.2 Recovery of pure compounds from spectroscopic examination samples

Type of spectroscopic examination	*Matrix of sample*	*Recovery method for obtaining pure sample*
UV, visible spectra	Solvent	Remove solvent by evaporation
IR spectra	Liquid paraffin	Dissolve mull in petroleum ether, pass through a silica column. Wash with petroleum ether, and then elute sample with a more polar solvent
IR spectra	KBr disc	Crush disc in mortar, dissolve sample in appropriate non-aqueous solvent, filter, remove solvent by evaporation
NMR spectra	Solvents containing internal standard, e.g. TMS	Use preparative TLC or remove TMS by using short clean-up column of silica gel (TMS will be eluted by very non-polar solvents)

TMS, trimethylsilane.

The amount of compound to be used for physico-chemical investigation is about the same for all compounds with molecular mass less than 2000 and is shown in Table 8.3. UV–VIS and NMR spectra are usually determined when the sample is in solution, and this information is also important for obtaining mass spectra and for applying materials to chromatographic systems. Relatively non-polar solvents are most commonly used since these interfere less with the detection systems and are more volatile (Table 8.4).

Table 8.3 Amounts of compound needed for various physico-chemical measurements

Technique	*Amount needed (mg)*
UV/visible spectrophotometry	2
IR spectrophotometry	
KBr disc	2
paraffin mull	10
Mass spectrometry	1
^{1}H NMR spectroscopy (>200 MHz)	10
^{13}C NMR spectroscopy	20
	(more for correlation spectra)
Analytical chromatography	
TLC	2
GC	1
HPLC	2

Table 8.4 Solvents (in order of preference) for physico-chemical measurements

Physico-chemical measurement	*Common solvents*
UV/visible	Ethanol, methanol (water for visible)
IR	Chloroform
NMR	Deuterochloroform, D-6 DMSO or D-5 pyridine (for polar compounds), D-6 acetone, D-4 methanol
Mass spectrometry	Chloroform, methanol

8.3 QUANTITATIVE ANALYTICAL METHODS

Quantitative analysis involves the measurement of the amount of one or more substances in a sample. A simple quantitative analysis that should always be performed during extraction, fractionation and isolation is to weigh the sample at each stage from the original organism to the isolated compound so that the yield of the material can be calculated and recorded. This is particularly important during preparative chromatography as it will give an indication of how much material is being lost by, for example, binding to the stationary phase. Analyses based on weight are known as gravimetric analyses.

Other quantitative methods that can be applied to crude or purified natural products are titrimetry, quantitative spectroscopy and quantitative chromatography. The first two methods are less selective since the analytical signal is derived from a measurement performed on the entire sample. In many such cases the reading obtained may be due to a number of related chemicals sharing the property being detected in the test. In chromatography, however, the components of a mixture are separated and quantitative information on each individual component can be obtained by examination of the individual TLC zones or HPLC and GC peaks.

For any quantitative analytical protocol, the following parameters should be assessed before it is used for routine analysis: precision, accuracy, linearity, range, sensitivity, selectivity and robustness. The method should only be used under conditions in which the above parameters are achieved to an acceptable standard. A range of quantitative methods applicable to natural products in pure or mixed form are discussed further in this section.

8.3.1 Quantitative analysis using titration

Titration is a simple form of analysis which relies on a reaction occurring between the substance to be analysed (the analyte) and a reagent added to it (the titrant). This reaction can be of different types, such as acid–base neutralization, metal–ligand complexation, precipitation or oxidation–reduction. The general method is:

1. Place an accurately weighed (to four decimal places) amount or accurately measured volume (if a solution) of the sample in a conical flask.
2. Dissolve in a suitable solvent if necessary.
3. Add a small quantity of an indicator (a substance which gives a colour or other detectable change when the reaction is complete).
4. Add the titrant in small, accurately measured increments (a burette is usually used) until the indicator changes colour – the 'end point'.
5. Note the total volume of titrant added and calculate the amount of analyte in the sample. The stoichiometric relationship (i.e. the number of molecules of titrant reacting with each molecule of analyte) between the analyte and titrant must be known.
6. The titration is repeated in duplicate or triplicate. The analysis is valid if the difference between replicates is less than 1% of the mean reading. A larger variation may indicate either poor technique or the occurrence of degradation reactions during the titration.

Acid–base titrations can be used to measure the alkaloidal content of an extract or the fatty acid content in an oil or wax. 'Back titration' methods are often used, where excess reagent is added to the sample, and the unreacted residue of the reagent estimated by titration with a suitable titrant. In the analysis of alkaloids for example, a crude alkaloidal extract in free base form may be treated with excess hydrochloric acid. Unreacted acid is then measured by titration with sodium hydroxide. This method has the advantage of giving well-defined titration curves (since HCl and NaOH are a strong acid and base, respectively) whereas the end point from a direct titration of the alkaloid (a weak base) with acid would be less precise. Titrations involving addition of iodine (as iodine monochloride) to double bonds in fatty acids, followed by estimation of excess iodine by sodium thiosulphate are used in measuring the degree of unsaturation of oils and waxes.

Titration methods are relatively simple to perform and require no sophisticated equipment. A limitation is that when applied to a sample consisting of a mixture, titration is a non-selective method and tends to quantify a group of related substances sharing a common feature, e.g. basicity, acidity or unsaturation, rather than a single member of a group. Since the molecular weights of each member of the chemical group will be different, that of the predominant substance may be used for the purposes of the calculation, necessitating the use of phraseology such as 'the alkaloidal content was 1.35% w/w calculated as reserpine'. However, if the sample is an isolated compound, then titration can be valuable in assessing its percentage purity since the exact volume of titrant required for a 100% pure sample can be calculated, deviations indicating the presence of contaminants.

8.3.2 Quantitative analysis using spectroscopy

All quantitative spectroscopy relies on the Beer–Lambert law, which states that the absorbance of light (*A*) by a particular substance is related to its concentration (*c*) in the sample and the path length (*d*) through which the light travels. Absorbance is defined as the $\log_{10}$ of the ratio of intensities of light entering (I_0) and emerging from (I) the sample.
Thus:

$$A = ecd$$

where *A*, *c* and *d* are as defined above and *e* is a proportionality constant which depends on the sample, the solvent and the wavelength of measurement. Rearranging the equation, we get:

$$e = A / cd$$

where the *e* is the absorbance per unit concentration and unit path length. The constant *e* is called the extinction coefficient and two forms are commonly encountered. These are the molar extinction coefficient (ε_M) – the theoretical absorbance of a 1 M solution with 1 cm path length – and the *A* (1%, 1 cm) value, which is the theoretical absorbance of a 1% w/v solution with 1 cm path length. Absorbance readings are taken in a UV or UV–VIS spectrophotometer, or a colorimeter (visible only).

The Beer–Lambert law applies in both the UV (190–340 nm) and visible (340–800 nm) regions of the electromagnetic spectrum, and can be utilized in the quantitation of substances. If the extinction coefficient of a substance in a particular solvent and at defined wavelength is known, then by measuring the absorbance of a solution and using the equation $A = ecd$, the concentration of the analyte can be calculated since c is the only unknown.

In the absence of a value for extinction coefficient, two approaches can be undertaken. One is to prepare a standard solution of known concentration using authentic, pure, reference material of the substance to be analysed. The extinction coefficient can be calculated from the standard solution and applied to the Beer–Lambert equation to calculate the concentration of the sample solution. The second, more accurate method is to construct a calibration curve in which a series of solutions of the reference material are made and the absorbance of each is measured. A graph called a calibration curve (Figure 8.6) is plotted using absorbance (*y*-axis) as the signal against concentration (dose, *x*-axis), from which the concentration corresponding to the measured absorbance of a sample can be read off. The concentration range used to prepare the calibration should be from at least 50 to 150% of the anticipated analyte reading. A linear graph is expected.

Some practical points to note are:

1. Measurements should always be taken against a blank consisting of the solvent in a closely matched cell.
2. Plastic or glass cells may be used for measurements in the visible region, but only quartz cells for the UV region since the former materials absorb UV light.
3. The solvents used to prepare the solutions, the wavelength at which absorbance is measured, and the path length should be the same for the reference standards and the samples.
4. Volatile solvents should be avoided or used only in capped cells since evaporation of the solvent during measurements can lead to errors.
5. The wavelength of choice for quantitative spectroscopy is that corresponding to the maximum absorbance in the spectrum (λ_{max}). This has the advantage of being a characteristic feature of the analyte, producing the largest reading for a given concentration and causing less inaccuracy if a slight error in wavelength setting occurs (due to the flat top at the peak).
6. Absorbance readings above a value of 2 ($I < 1\% \, I_0$) are not likely to be accurate (unless a highly accurate modern instrument is used), and the

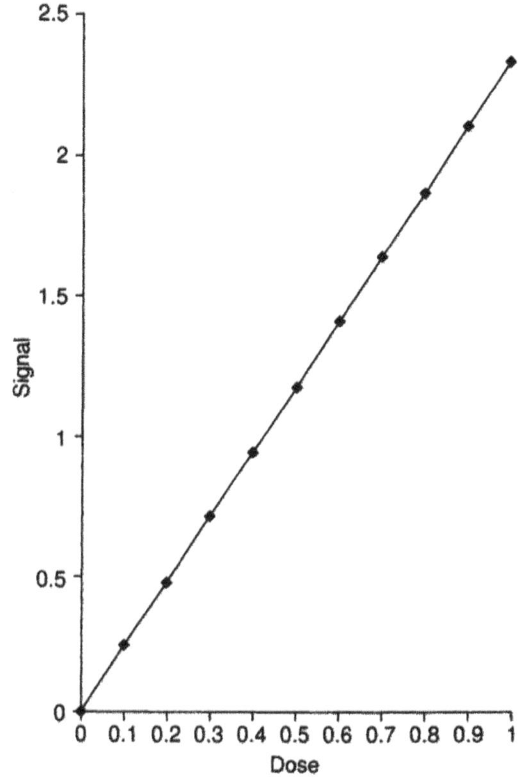

Figure 8.6 Calibration curve.

concentration should always be adjusted to give readings less than this value.

For samples that absorb UV light or are coloured, direct measurement of absorbance at a suitable wavelength is possible. However, due to the presence of a large number of UV or visible light-absorbing materials in extracts, some clean-up prior to measurement is necessary. In some cases it may be useful to prepare a coloured or UV-absorbing derivative. The derivatizing reaction used may be specific to a particular chemical group (section 8.2.1), for example anthraquinones may be reacted with alkali to form a pink colour and their concentration quantified by measurement of the colour intensity. The Kedde reagent is used to make a quantifiable coloured derivative from cardenolides, iron (III) chloride forms blue complexes with phenolic compounds and the antimony trichloride reaction of carotenoids can be adapted for quantitative purposes. In each case, a standard amount of reagent (alkali, Kedde reagent, iron (III) chloride, or antimony trichloride in the above examples) is added to different concentrations of a standard reference material, and to the sample under test.

Colour intensity (absorbance) of the reference derivatives is plotted against their concentration (as in Figure 8.6) and the concentration of the sample obtained from the absorbance value of its coloured derivative. Non-linearity in the upper parts of the calibration curve most probably indicates that there is insufficient derivatizing agent for full reaction with higher concentrations of the reference material.

The advantage of this method is that it can result in selective measurement of a single group of chemicals present in an extract or fraction. However, it is seldom selective for individual members of a group and so the total concentration of all substances reacting with the reagent will be obtained. The reference material will usually be a single compound, perhaps identical with only one of those present in the mixture (e.g. digoxin is used in the *British Pharmacopoeia* in a test for total cardenolides) or a synthetic compound known to give the appropriate colour reaction may even be used (e.g. 1, 8-dihydroxyanthraquinone is used as a calibration standard for the measurement of *Frangula* anthraquinones). Thus the actual concentration value obtained is only approximate, and will need to be recorded as '0.5% w/w cardenolides calculated as digoxin' or '6% glucofrangulins calculated as 1,8-dihydroxyanthraquinone', etc.

Where spectrophotometers are not available, visual inspection may be used to estimate concentrations of analytes in extracts by comparison of colours of sample solutions (occurring before or after derivatization) with those of standard solutions.

8.3.3 Quantitative analysis using thin-layer chromatography

Chromatographic methods for quantitative analysis offer the additional advantage of separation of the components of a mixture, thus allowing their individual quantitation. GC and HPLC both give simultaneous qualitative and accurate quantitative measurement and are used extensively (section 8.3.4). However, these two techniques require costly apparatus and involve high running costs. On the other hand, the low cost, rapidity and good resolution associated with thin-layer techniques have meant that these techniques are often used for routine screening procedures, even though they are not as accurate as the more sophisticated methods.

In a TLC chromatogram, the size and colour intensity of the zones are related to the amount of substance present. This can be used for quantitative analysis. The technique is essentially the same as that used for analytical TLC, with the following additional points to be borne in mind:

1. The amount of substance initially applied to the plate must be known accurately. This is usually ensured by applying an accurately measured volume of a solution of known concentration. Graduated capillary pipettes or a microsyringe can be used to measure the

volume spotted on to the plate. Loss of accuracy due to variations in the volume applied may be overcome by using an internal standard (section 8.3.4).

2. To ensure reproducibility of any chromatographic method used quantitatively it is necessary to adhere to the specified conditions of, for example, temperature, humidity, solvent and adsorbent. In this way, some degree of reproducibility can be attained.
3. The zone produced after development must exhibit a measurable function proportional to the amount of substance present. There are three major ways of measuring this relationship:
   (a) Visual inspection. Examining the size and intensity of the sample zone in relation to those of the standards applied will give an approximate idea of the concentration of the sample.
   (b) Elution of the zone and subsequent spectrophotometry. This technique involves the removal of the stationary phase containing the zone from the plate, elution of the compound from the adsorbent with a suitable solvent, and measurement of absorbance (usually with UV light, or a coloured derivative is made and visible light used). The amount present can be calculated using the Beer–Lambert law (section 8.3.2).
   (c) The intensity of the zone, i.e. its 'colour' *in situ*. The higher the concentration of substance in the zone the greater is its intensity, as can be seen if the compound itself is coloured or a coloured derivative is made on spraying. The intensity of zones can be measured by a densitometer. This is an instrument which, by means of an optical system and photoelectric cells, measures the amount of light either reflected from or transmitted through a zone relative to that impinging on the background adsorbent. The plate is moved slowly at a constant speed under the scanner and a differential curve is produced on chart paper. The area under each peak is proportional to the intensity of the zone and hence the amount of substance present. Peak area can be obtained electronically using an integrator or can be calculated manually by treating each peak as a triangle and using the formula:

   $$\text{area} = 0.5 \times \text{base} \times \text{height}.$$

   A correction factor of 0.94 can be applied to allow for the absence of an acute apex. An alternative formula is to multiply the height of the peak by its width at half-height (Figure 8.7). Recently developed densitometers can give quantitative information of high precision and accuracy, provided the initial chromatogram was generated in a suitable manner. These instruments are relatively costly. An inexpensive modification of the densitometer concept is to use a computerized scanner. The scanner scans the plate and displays an image on the computer screen. Suitable software can

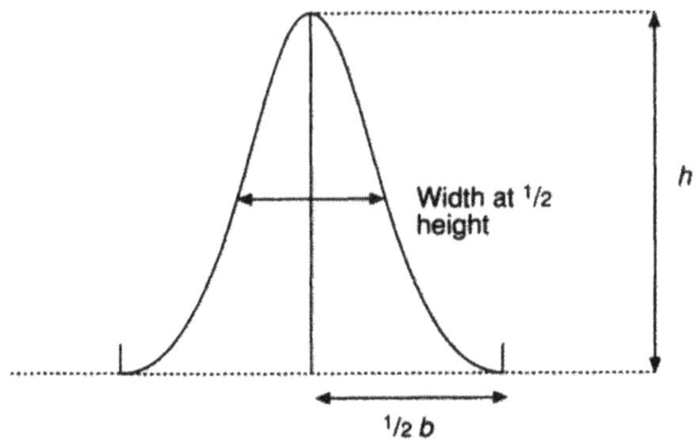

Figure 8.7 Measurement of peak area.

be used to convert the intensity or darkness of each zone in the image to a peak, or to give a numerical readout corresponding to the intensity of the zone. The success of this method depends on the contrast between the zones and the background.

Whichever of the above three methods is used, known amounts of a standard substance must be chromatographed alongside the unknown. The amount of substance present in a spot from an unknown sample is usually determined by reference to a calibration curve obtained by measuring zones of known amounts (Figure 8.6). Preferably, these zones should be on the same plate as the test mixture in order to minimize errors caused by variations in the chromatographic and detection conditions. It is desirable that a separate calibration be prepared for each component of a mixture to be quantified, since the response to detection reagent and the exact relationship of zone size and intensity to amount will vary for each substance.

8.3.4 Quantitative analysis using GC and HPLC

GC and HPLC are commonly used for the quantitative analysis of substances due to the high sensitivity, resolution, precision and accuracy of these methods. The parameter measured for quantitation of a substance is the area of its peak in the chromatogram, which is proportional to the amount of that component injected. Thus a simple calibration would be to inject known amounts of a reference standard, plot a graph of peak area versus amount (or concentration if constant volumes are injected) (Figure 8.6), and use this to quantify analytes in sample solutions. This method can be used fairly successfully in HPLC where loop injectors allow the delivery of constant volumes with a high degree of precision. Thus a series of concentrations of the reference standard are made up, equal volumes of each are injected and a calibration curve prepared of peak area versus concentration.

This method is not so suitable, however, for GC, where injection of a sample is usually carried out manually by means of a syringe and needle. The volumes are also smaller (section 8.2.2) leading to greater inaccuracies. To overcome such errors, an additional substance, or internal standard can be added in constant amount to all standard solutions and to the samples. For the calibration, a graph is plotted of peak area ratio (analyte/internal standard) versus concentration ratio (analyte/internal standard). The peak area ratio determined for a sample is related, using the calibration curve, to a concentration ratio. Since the concentration of the internal standard in the sample is known, the concentration of the analyte in the sample can be calculated.

The characteristics of an ideal internal standard are that it must:

- not co-elute with any of the components in the sample;
- not be a substance normally or sometimes found in the sample;
- not react chemically or physically with any of the components in the sample;
- have a reasonable retention time in the chromatographic system used;
- be stable under the chromatographic conditions used;
- be detectable by the detection method used;
- not evaporate from solutions of the sample;
- not precipitate from solutions of the sample; and
- be used at a concentration giving similar peak size to the analyte peaks.

Many analyses using GC or HPLC routinely employ the use of an internal standard. This may be added just prior to chromatographic analysis, but may also be added earlier on in the work-up processes, particularly if sample losses are expected, e.g. where solvent : solvent extraction is involved and the two phases cannot be separated fully or reproducibly. If the internal standard is added in the early stages of a process, it is important to ensure that it is not lost due to instability, poor extraction or evaporation. When using the internal standard method, it is good practice to run a chromatogram of the sample without any added internal standard to ensure that there are no components present in the sample which are eluting at the same retention time as the internal standard, thus giving a false reading for the peak area of the latter.

8.4 BIOLOGICAL ASSAYS

8.4.1 Dose of fraction to be used

A very wide range of biological assays may be performed on substances derived from natural sources. These range from tests on simple isolated enzyme systems and cell-based models to more complex tissue and animal models, and ultimately clinical trials in humans. The amount of substance needed for biological assessments, i.e. the dose, will vary considerably according to the technique in question and how many experiments are to be carried out. Ideally a range of concentrations should be used and each determination repeated at least five times for statistical validity, so it is necessary to bear such facts in mind when calculating the amount of a fraction needed for an experiment to be meaningful. The advent of small-scale bioassays using microtitre plates has considerably reduced the amount of extract that needs to be available for use.

Once an effective dose for the total extract has been determined, the dose range to be used at each fractionation stage should reflect the dilution involved. A parameter such as the EC_{50} (the dose required to give 50% of the maximum response) or IC_{50} (the dose required to produce 50% inhibition of the control value) may be determined to compare the activity of extracts or isolates (Figure 8.8). The EC_{50} of active fractions should decrease at each fractionation stage, as the fractions become more and more enriched with the active constituents, and the final EC_{50} of major active components is expected to correspond roughly to the equivalent amount calculated to be present by assay in the original extract. However, the EC_{50} of the isolated constituent may be higher than expected if synergism or additive effects occur in the unpurified extract, or if considerable decomposition has taken place during the fractionation process.

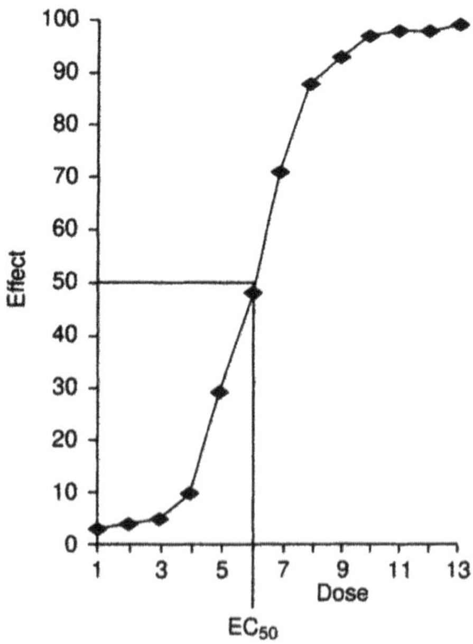

Figure 8.8 Dose–response curve to calculate EC_{50}.

8.4.2 Solubility of fractions or isolated compounds in the bioassay medium

Unlike many of the spectroscopic methods used for structure elucidation which require samples to be dissolved in organic solvents, most *in vitro* biological tests utilize an aqueous environment since this conforms to the *in vivo* state. There is thus some conflict between the desirable solubility features of a fraction, depending on the tests that are intended, and some skill is needed in presenting extracts and compounds in the forms that will yield the most meaningful results. In every case it is desirable for all the fraction to be completely dissolved in an appropriate solvent. If a biological test is to be performed, care should be taken that the solvent giving best solubility is compatible with the system. Many non-aqueous solvents are toxic, especially to cultured cells and to tissue systems. Tests to determine the concentration of solvent above which toxicity occurs should always be carried out before the experiment proper, and controls with potential solvent toxicity in mind should be incorporated into the experiment.

The most common problem experienced in biological testing is the preparation of a solution of a low-polarity fraction in an aqueous medium. Ethanol and dimethylsulphoxide (DMSO) are the two most common solvents since they are miscible with water and, at concentrations below 3% v/v, are usually not toxic. Care should be taken that precipitation does not occur when the solution of the fraction in the solvent is diluted with water to the desired level. If no suitable solvent can be found, then emulsions and similar preparations can be made using non-ionic emulsifiers such as Tween®, but the toxicity of these at the concentrations used should always be checked.

8.4.3 Sterility

In any biological test where cells and micro-organisms are cultured, it is vital that all media and solutions used are sterile since bacterial and fungal contaminants may suppress the growth of cells used for the test system, produce false results and interfere with the readings to be taken. Preparations of fractions, their dilutions and all tests should therefore be carried out under aseptic conditions, preferably in a laminar flow cabinet.

The production of a sterile extract presents some problems, especially if aqueous solvents are used for making or diluting the extracts, since even if the water is sterile, it is unlikely that the source material will be. The most common sterilization procedures use high temperatures or high doses of radiation, and both of these rather drastic measures may cause decomposition of the active constituents. Consequently, sterilization is often achieved by filtration of the extract (pore size < 0.4 µm) once it has been made. Alternatively, the sample can be dissolved in a water-miscible solvent, such as 100% DMSO, ethanol or methanol, which does not support bacterial growth and will be sterile. Subsequent dilutions are carried out under aseptic conditions using sterile aqueous media, ensuring that the final concentration of the organic solvents is not likely to interfere with the bioassay. Fractions that are to be dissolved ultimately in dilute alcohol solutions can be sterilized by exposing them to the correct amount of 100% alcohol for at least 1 hour before dilution to the correct concentration.

8.5 BIBLIOGRAPHY

Braithwaite, A. and Smith, F.J. (1985) *Chromatographic Methods*, 4th edn, Chapman & Hall, London.

Dappen, R., Arm, H. and Meyer, V.R. (1986) Applications and limitations of commercially available chiral stationary phases for high performace liquid chromatography. *Journal of Chromatography*, **373**, 1–20.

Dey, P.M. and Harborne, J.B. (series eds) *Methods in Plant Biochemistry*, Academic, London.

Fifield, F.W. and Kealey, D. (1990) *Principles and Practice of Analytical Chemistry*, 3rd edn, Blackie, Glasgow.

Harborne, J.B. (1984) *Phytochemical Methods*, 2nd edn, Chapman & Hall, London.

Ikan, R. (1991) *Natural Products – A Laboratory Guide*, 2nd edn, Academic Press, London.

Williams, D.H. and Fleming, I. (1991) *Spectroscopic Methods in Organic Chemistry*, 4th edn, McGraw-Hill, London.

CHAPTER NINE

General methods

The methods used in the extraction, fractionation and isolation of compounds from natural sources have been described in Chapters 3–7. A knowledge of certain basic general techniques is useful, and these are described in more detail here.

9.1 BASIC TECHNIQUES IN HANDLING EXTRACTS

In the extraction of organic material followed by its clean-up and fractionation, it is usually necessary to separate a solid residue from a liquid by filtration, and in many cases to remove solvent by evaporation. Some methods for this are described here.

9.1.1 Filtration

Filtration removes solid particles from a liquid in which they are suspended by preventing their passage through a filter consisting of many small holes in a rigid membrane. The size of the holes (called pores) obviously determines the size of particles that are retained, and the more expensive filtration media contain pores where the size is determined accurately.

A wide range of pore sizes (usually denoted by the diameter of the hole) in the membranes used for filtration exists and this makes possible a variety of applications. Coarse filters (those with large pores) are used for straining extracts and removing plant or other solid organic material. Finer filters are used for removing precipitates, whereas the smallest sizes can used to remove micro-organisms and thus sterilize solutions. Filters with very small pores often need the application of positive or negative pressure to enable a fast flow rate of liquid through the filter to be achieved.

Types of filter

The matrix forming the filter can be made of a variety of materials. The most basic and the cheapest material is paper. This is suitable for many purposes but has the disadvantage that the cellulose fibres, of which the paper consists, swell in the presence of water, especially if it is alkaline. The result of this is a decrease in pore size and consequent reduction in flow rate. Another major disadvantage is that aqueous solutions are absorbed by the fibres and so a significant amount of extract or fraction will be lost unless the paper is washed thoroughly with the solvent used after the extract is passed through.

Small quantities can be filtered using a funnel and the filter paper folded into a pleated cone. It is, however, much faster and more efficient to filter using reduced pressure and a Buchner funnel, as shown in Figure 9.1. This has a flat ceramic or glass bed on which the filter paper is laid. The liquid to be filtered is then sucked through.

Special papers are available which are treated with silicone so that they do not permit the passage of aqueous solvents but do enable 'organic' solvents to pass through. They are thus useful for removing small amounts of water and resolving emulsions.

Sheets similar in appearance and form to paper can also be made of glass fibre. They have the advantage that they do not swell in the presence of water but they are more fragile and not very flexible. Sintered glass filters are commonly used as a more robust form of filtration through glass. These consist of a porous glass matrix several millimetres thick which is fused into a funnel. The pore size can be controlled as part of the process of making the sintered bed, so a range of different filter sizes is available.

Overcoming blockage of filter

A common problem that occurs in filtration of suspensions with a large number of fine particles is that the pores become blocked and an almost

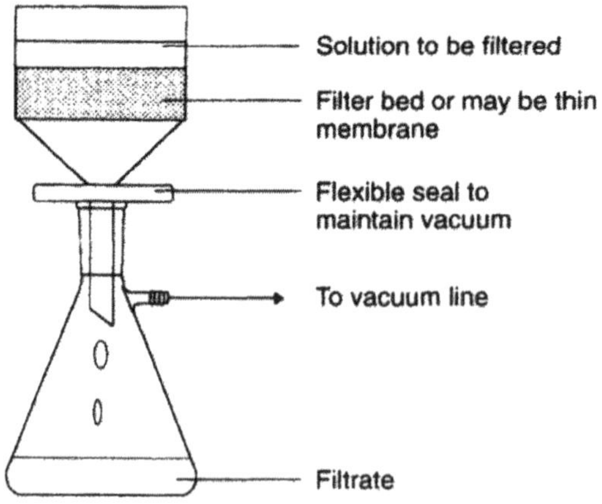

Figure 9.1 Buchner filtration system.

impermeable layer of the fine particles forms, with a corresponding drop in flow rate and the result that the filtration process will take a very long time. The filtration rate can be improved by careful disturbance of the layer with the flat end of a blunt spatula and by continuously stirring the suspension. It is very easy to damage the filter membrane and so this method is not very satisfactory. It is preferable to make a thick filter bed with a porous material such as kieselguhr or very fine sand:

1. Set up a Buchner filtration system using a coarse sintered glass filter, paper or glass fibre paper.
2. Measure out enough kieselguhr for filtering (or fine washed sand) to fill about a third of the funnel.
3. Add half of the kieselguhr to the solvent which has been used in the extract or suspension which is to be filtered.
4. Stir the kieselguhr/solvent mix until an even suspension is formed.
5. Moisten the filter paper or bed with the solvent to be used. Apply negative pressure so that the sheet is sucked against the support in the funnel. **It is important to ensure that there are no wrinkles or air spaces around the edge of the paper**.
6. Stir the kieselguhr/solvent mix and carefully pour on to the filter bed with negative pressure still being applied. If necessary, do not pour in one place but move the lip of the beaker over the surface so that a bed of even thickness is formed.
7. Wash any remaining solid from the beaker with the solvent and pour through the layer. Maintain a moderately low negative pressure and do not allow the bed to dry out and crack.
8. Discard the filtrate from this operation.
9. Treat the solution to be filtered with the remaining kieselguhr as in (4) and pour on to the bed. Maintain negative pressure and collect the filtrate.
10. Wash the bed with the solvent alone and add the filtrate to the filtrate obtained in (9).
11. If required, elute the substances retained on the filter bed as follows:
    (a) Release the negative pressure.
    (b) Add a solvent to the bed which will dissolve the compounds of interest.
    (c) After at least 30 min, apply the negative pressure and collect the filtrate. This should now contain in solution the substance(s) originally retained on the layer.

9.1.2 Evaporation of solvent

On many occasions during fractionation procedures it is necessary to concentrate a relatively large volume of solvent to a much more concentrated form or even to a dry residue. This is in order to concentrate the extract for chemical or biological analysis, to remove any solvent traces which may interfere with the assay, or to allow the dry residue to be weighed and stored or reconstituted in a different solvent for further processing. It is important that removal of solvent is carried out at as low a temperature as possible to avoid decomposition of constituents.

The method chosen for removal of solvent will depend on the volume of the extract, volatility of the solvent, the thermostability of the extracted substances and the speed with which evaporation is required to be carried out. It is advisable in all cases to preweigh the empty container in which final evaporation is to take place so that the final yield can be easily calculated without the need to transfer the extract to another container (with the risk of losses). It is also advisable to dry extracts to constant weight to confirm that complete removal of solvents has taken place. Methods available are as follows.

Simple evaporation by heating

Place the extract in an evaporating dish either at room temperature or over a heating aid (e.g. a water bath). This is best carried out in a fume cupboard so that evaporating solvent vapours are led away into an exhaust. CARE!!! Flames should not be used to heat evaporating dishes containing organic solvents. This method is suitable for only small-scale extractions.

Evaporation under reduced pressure

The extract is placed in a round-bottomed flask and the solvent removed using a rotary evaporator (Figure 9.2). Most solvents can be removed fairly quickly if they are evaporated under reduced pressure. By this method the boiling point of the liquid is reduced so decomposition is minimized and the rate of evaporation at ambient temperature increases.

Reduction of pressure is attained by an attachment to a Venturi apparatus on a water tap or to a vacuum pump. It is important to include a water trap on line in the first instance to prevent any suck-back into the flask containing the solution to be evaporated. The solvent collects in a vessel placed under the condenser and can be recycled (if a single solvent) or disposed of in an appropriate manner. Heat may be applied by means of a water bath to assist evaporation, provided the substances extracted are not thermolabile. Extracts should be transferred to sample tubes for final storage.

This method has a number of limitations. It may result in the loss of volatiles from the extract, particularly where heat is applied. The apparatus must also be kept scrupulously clean to avoid contamination by trace residues from other extracts for which it has been used. It is difficult to concentrate aqueous solvents unless a very low pressure can be obtained. Even so, high temperatures have to be applied to the solution

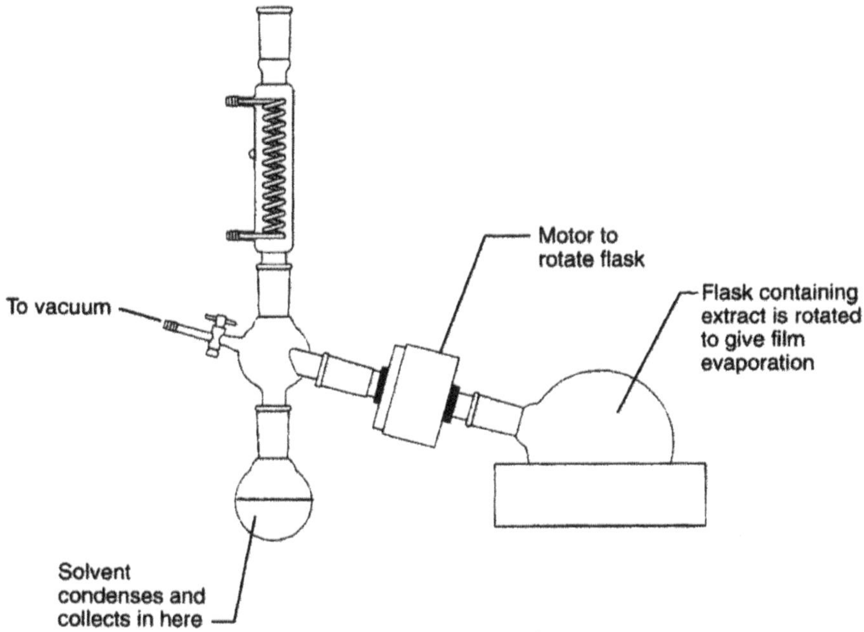

Figure 9.2 Rotary evaporator.

to facilitate rapid evaporation. Sudden boiling can occur easily, with the result of some liquid being lost from the flask. This can be prevented by using small amounts of solution and adding some anti-bumping granules to the solution.

The use of reduced pressure may pose an additional problem for aqueous extracts containing saponins as considerable frothing is observed in this case. The froth may travel out of the sample flask and into the main body of the rotary evaporator. Frothing may be controlled by constant vigilance and careful control of the vacuum (releasing it whenever frothing begins) or the addition of a few millilitres of propanol or butanol to the extract. This boils off along with the water, but will need to be replaced periodically during evaporation of the extract since the alcohol will boil off more readily.

Evaporation with gas stream

Small amounts of volatile solvents can be evaporated by blowing a steady stream of an inert gas, usually nitrogen, over their surface. This is often useful at the last stages of fractionation when only a few milligrams of sample may be produced from the solution and the solution is easily contained in a small, preweighed tube which is its final container.

A variation of this method is the use of a high-speed centrifugal chamber where the samples are spun round at a high rate of revolution. In this case the atmosphere consists either of air (thereby raising the risk of oxidative decomposition of mixture components) or of nitrogen.

Freeze-drying

Freeze-drying may be used for aqueous samples, and generally gives a drier and less sticky residue than the methods described above. A variety of freeze-driers is available commercially. The technique is particularly useful for proteins and other thermolabile, water-soluble samples.

In this method, the sample is prefrozen and placed in a chamber evacuated by a vacuum pump. The water 'sublimes', i.e. goes straight from ice to vapour phase and is collected on a condenser. In most cases the dry residue is a fluffy solid. To prevent the dry residue from being sucked out of the container by the vacuum, the frozen extract should be covered with filter paper held in place by a rubber band before placing in the freeze-drier. Alternatively 'cling-film' may be used and a series of small holes made in the film before the sample is placed in the freeze-drier. A wide-mouthed container should be used to ensure efficient drying of the extract. Beakers or porcelain evaporating dishes may be used, but the authors have obtained good results using non-stick baking trays with a depth of about 3 cm. Freeze-drying should never be used where there are traces of organic solvent or volatile acids or bases in the extract as this will cause deterioration of any rubber tubing and damage to the pump by dissolving in the lubricating oil. In any case, a mixture of water with another solvent such as ethanol has a much lower freezing point and may not be in the solid state required for the technique.

Vacuum dessicators and ovens

Final drying of samples that appear dry, e.g. for spectroscopy, can be achieved with the help of a vacuum dessicator or vacuum oven. This method is not suitable for the removal of visible quantities of solvent.

Due to the sticky nature (even when dry) of many of the extracts obtained using simple evaporation or evaporation under reduced pressure, it is advisable to evaporate most of the solvent using these methods in a larger container and then transfer the concentrated solution to a preweighed sample tube and to remove the last traces of solvent by freeze-drying or evaporation with a gas stream.

For an aqueous extract that cannot be evaporated straight away, it is advisable to store it in a frozen state. This will prevent any breakdown due to enzymes (e.g. glycosidases) present in the extract and will also minimize colonization by micro-organisms.

9.2 ISOLATION OF ACTIVE COMPOUNDS

When one or more fractions has been obtained which seems to consist of one major compound when analysed by chromatography, the aim must be to isolate that compound in a pure state so that its structure can be determined and its dose–activity profile determined in the biological system in question. This section is concerned with ways in which this can be achieved.

9.2.1 Crystallization

If the fraction is concentrated and left to cool, and the solvent allowed to evaporate slowly, then crystals may form of the pure compound. Crystallization can be encouraged by slight scratching of the inside walls of the glass with a glass rod and by leaving in a cool place, even in a refrigerator.

Any deposit may be crystalline and should be checked with the aid of a hand lens to make sure that it is not a deposit of amorphous material coming out of solution as cooling has taken place.

Even if the deposit appears crystalline, it should be checked chromatographically for purity as described below, since closely related substances can co-crystallize under certain conditions.

The deposited crystals should be separated from their mother liquor by filtration or centrifugation and then redissolved in a small amount of the fresh solvent and recrystallized.

The melting point and optical rotation of the crystals should be determined once it is confirmed that they are pure. In some cases, if the crystals are large enough, X-ray diffraction studies can be carried out.

In some cases, the compound crystallizes more easily as a derivative than as the parent compound. Thus alkaloids in the free base form are often difficult to crystallize whereas their salts crystallize comparatively easily. However, it is important to realize that, although this is useful for structural determination purposes, it is not suitable for biological activity studies since the two compounds may have different properties.

Because of the small amounts of pure material eventually isolated, crystallization is often not achieved in the isolation of naturally occurring compounds. If this is the case, then isolation of single substances and determination of their purity is achieved by chromatography.

9.2.2 Use of chromatography

For isolation of single substances

The most common method of isolation of pure substances at the final stage is to use preparative TLC (section 7.2.2). In some cases where TLC

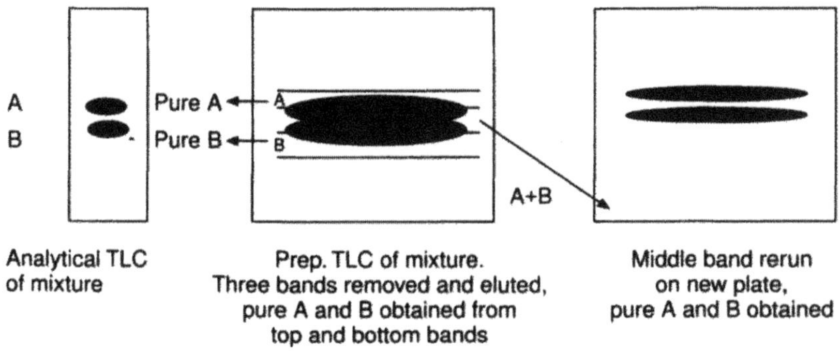

Figure 9.3 'Cut and rerun' process for isolation by preparative TLC.

does not give adequate resolution, preparative HPLC can be used, but the former method is less expensive and amenable to low-budget conditions.

If two compounds run very close together and are not fully resolved by the separation process, a two-stage 'cut and rerun' process may be necessary, as shown in Figure 9.3.

For checking purity

The purity of isolated substances should always be checked by at least four different TLC systems, using different mobile phases of about the same polarity. If possible, HPLC should be carried out and the symmetry noted of the peak produced. Once the identity of the compound has been suspected, this should be confirmed by 'spiking' the substance with authentic material and noting that no broadening or formation of shoulders occurs in the peak profile.

9.3 CHROMATOGRAPHIC PROCEDURES

9.3.1 Packing chromatographic columns

Columns can be either dry packed or wet packed.

Dry packing

In dry packing the solid stationary phase is introduced into the column and allowed to settle. This can be aided by gently tapping the bottom end of the column on a hard surface and also tapping the walls with a somewhat flexible rod. The sample (section 6.1) is applied to the column, allowed to settle as an even band and the mobile phase is then allowed to pass through the column. It eventually reaches the end of the column and passes out into a collection device.

It is difficult to achieve a narrow band of applied sample with dry packing, and it is difficult to achieve uniform flow of the mobile phase because it is difficult to attain even packing of the stationary phase. Air bubbles are often trapped as the mobile phase passes through and this further disrupts the flow.

Wet packing

Wet packing is preferable and is recommended here as the method of choice.

1. Measure enough stationary phase to fill about half of the column. The easiest way to do this is to half-fill the dry column with the relevant dry powder, tap it to allow some settlement and add a little more if necessary.
2. Pour the stationary phase out of the column and weigh it. It is important to know the weight since this gives an indication of the amount of extract that can be used so that the column is not overloaded. No more than 1 g total extract per 100 g stationary phase should be used. On the other hand, it is important to minimize loss by irreversible adsorption to the stationary phase so not less than 400 mg total extract per 100 g stationary phase should be used.
3. Suspend the stationary phase in the mobile phase chosen or one of lower polarity (or, if a reverse-phase system is being used, one of higher polarity) and stir gently until an even suspension is formed. The suspension should give a little resistance to stirring but it should be easy to pour. If it is too thick and viscous, air bubbles will become trapped and these will cause flow through the column to become disrupted. Silica gel with some solvents, such as chloroform, forms an almost transparent gel in which air bubbles can be seen easily, but other mixtures have the appearance of thick milk and it is much more difficult to see if any bubbles have formed. If an ultrasonic bath is available its use for a short time (<5 min) may help to remove all gases from the suspension.
4. Clamp or rigidly suspend the column in an upright position. Ensure that the tap at the bottom of the column is closed or that flow out of the end is not possible. This prevents the column becoming 'dry' and avoids the subsequent formation of air gaps and uneven flow once the suspension is poured into it.
5. Pour the suspension carefully and slowly into the column, tapping the wall of the column gently with a cushioned rod to encourage any air bubbles to rise to the top of the column bed. After a short time the suspension will settle and this can be encouraged by gentle tapping of the

bottom end of the column and its sides on a hard surface. Several centimetres of the mobile phase alone will eventually be seen above the surface of the stationary phase.

6. Carefully open the tap at the bottom of the column and allow the mobile phase to run out slowly until the height of the supernatant liquid above the column packing is less than 2 cm. Close the tap. It is very important not to let the level of mobile phase go below the surface of the stationary phase since, if this occurs, air will enter the bed, it will lose its even packing and the overall result will be very uneven flow of the mobile phase.

9.3.2 Preparation of the sample for application to the column

NB It is often best to prepare the sample and the column concurrently.

Introduction

The height of the sample applied to the beginning of a chromatographic system should be kept as small and as even as possible to achieve good separation. If it is broad, the bands will be even broader and will overlap as they travel through the column. If the applied mixture is not distributed regularly as an initial band, overlapping will again occur.

It is possible to apply the sample in liquid form, usually in solution, directly to the surface of the dry stationary phase. This leads to a very uneven starting band since the solution usually has to be dribbled over the whole surface by moving the pipette manually. The force of the fluid also tends to disturb the flat surface of the stationary phase, with the same result. This can be prevented by protecting the surface with a disc of filter paper or glass fibre or with a small amount of washed, clean sand. However, it is very difficult to ensure that the same volume has been applied to each unit area of the surface.

An alternative is to leave a small amount of the solvent used to suspend the stationary phase above the surface of the stationary phase. The sample liquid, which is then introduced, mixes with the supernatant and disperses quickly to form a uniform solution. This method achieves a much more even band but introduces some dilution of the sample and consequently a larger initial application.

In these two methods the sample has to be dissolved in the first mobile phase to be used. This may present difficulties, as generally a solvent of much lower polarity (higher if reversed-phase chromatography is being used) than that of the extract will be used as the initial eluant. A recommended alternative described here involves coating a small amount of the stationary phase with the sample and then introducing it to the top of the stationary phase in the column. This has the advantage that the sample can initially be dissolved in a solvent in which it is readily soluble.

Mixing sample with stationary phase

1. Determine the maximum weight of extract that the column will separate. 1 g extract for every 100 g dry stationary phase is a general guide.
2. Weigh out the correct amount of dry extract (previous evaporation or freeze-drying may be necessary at this stage), place in a porcelain or glass evaporating dish and dissolve it in not more than 2.5 ml of an appropriate solvent with appreciable volatility. Water can be used if freeze-drying facilities are available.
3. In a fume cupboard, mix not more than 2.5% of the stationary phase carefully with the extract solution until an even dispersion is formed. The more volatile solvents will evaporate quite quickly and the solution will become progressively more viscous and paste-like.
4. The extract becomes evenly coated on the stationary phase and eventually it dries and becomes a powder again. Less volatile solvents may need to be left overnight or for a similar period of time. If water is used to dissolve the sample, the paste of stationary phase and extract can be freeze-dried.

The sample is now ready for introduction to the column.

Introduction of the sample to the column

The sample can be introduced as a powder to the top of the column, either directly as a powder or as a suspension in the same solvent used to prepare the column. If the direct powder method is preferred:

1. Use a fine spatula to sprinkle the extract/stationary phase mix evenly over the surface of the supernatant in the column (not more than 2 cm depth supernatant). The extract will settle on the surface of the stationary phase.
2. It may be useful to sprinkle some more stationary phase into the supernatant once the sample has been added to prevent further dissolution of the extract into the mobile phase.

If the suspension method is preferred:

1. Suspend the extract/stationary phase mix in 2 ml of the same solvent used to prepare the column. Use a Pasteur pipette to transfer the suspension into the supernatant in the column, making sure that it is distributed over the whole surface.
2. Carefully open the tap at the end of the column and let the mobile phase run through until the level of the supernatant at the top of the column is just above (<3 mm) the top surface of the stationary phase (the top part of the stationary phase is derived from the sample and is often coloured to some extent).
3. Close the tap to prevent the bed drying out. It may be useful to

sprinkle some more stationary phase into the supernatant once the sample has been added to prevent further dissolution of the extract into the mobile phase.

4. Introduce the first aliquot of mobile phase eluant to the top of the column.
5. Open the tap at the bottom of the column and use it to adjust the flow. The optimum flow rate depends on the size of the column, the particle size of the stationary phase and any pressure applied to increase flow rate. The flow rate for a column 15 cm long and 2 cm diameter should not be greater than 1 ml/min for a conventional column and not more than 5 ml/min for a flash column.
6. Collect aliquot volumes of eluant. A fraction collector is most commonly used to do this and most modern machines can be programmed to collect at time intervals automatically. The aliquot volume is dependent on the size of the column and the number of components present which have similar retention behaviour. The maximum aliquot volume for a column 15 cm long and 2 cm diameter should be 5 ml.
7. The collected fractions should be monitored, e.g. by TLC or by a bioassay process, and fractions which have the same profile are bulked together. The bulked fractions should be taken to dryness and weighed.

9.4 BASIC TECHNIQUES IN THIN-LAYER CHROMATOGRAPHY

TLC is one of the most commonly used techniques in natural product research. This section deals with certain basic practical processes that are required in utilizing the technique.

9.4.1 Spreading a thin-layer plate

Although the best quality, robust, thin-layer plates can be bought ready-made in many countries, it is usually economically preferable to make plates in the laboratory which give a perfectly adequate separation. In some situations it may not be possible to buy the ready-made plates and 'home-made' ones are the only option.

The layer on such plates is more susceptible to damage than those available commercially but this can be minimized by careful handling. An additional benefit in making plates 'in house' is that the behaviour of the layer during the separation process can be changed by altering the fluid used in their preparation. The thicker-layer plates used for preparative work are usually 'home-made'. The stationary-phase material should be bought in bulk since it is difficult to make adequate materials in the laboratory.

Plates are made by mixing the finely powdered stationary phase with

Table 9.1 Quantities for making five 20 x 20 cm thin-layer plates

Size of plates	*Thickness of layer*	*Adsorbent*	*Suspending fluid*	*Special notes*
20 × 20 cm	250 μm	Silica gel 25 g	Water [a] 50 ml	Qualitative work
20 × 20 cm	1 mm	Silica gel 80 g	Water [a] 160 ml	Preparative work
10 × 6 cm	250 μm	Silica gel 10 g	Water [a] 20 ml	
20 × 20 cm	250 μm	Alumina 30 g	Water [a] 50 ml	Qualitative work
20 × 20 cm	1 mm	Alumina 100 g	Water [a] 160 ml	Preparative work
10 × 6 cm	250 μm	Alumina 12 g	Water [a] 20 ml	
20 × 20 cm	250 μm	Cellulose 40 g	Water [a] 60 ml	Mix in a blender before pouring
20 × 20 cm	250 μm	Polyamide 30 g	Acetone 45 ml	

[a] The nature of the layer can be altered by using a buffer solution to change the ambient pH or by using solutions of other salts to affect performance, e.g. 10% w/v silver chloride solution can effect separation of compounds differing only by a double bond.

water or another fluid to form a slightly viscous suspension. Suggested amounts are shown in Table 9.1.

The layer is formed by spreading the suspension as a thin layer on a run of about five glass plates by pouring into a trough which has a gap at the base of the trailing edge corresponding to the desired thickness of the plate. As the trough is pulled steadily along the line of plates, the suspension flows out from the base of the trough. The plates are left until the suspension dries and the plates are then removed, excess suspension removed from the two edges over which the trough moved and then heated in an oven at 105°C for at least 30 min before use.

CARE!!! Many of the fine powders used are dangerous if inhaled. A dust mask should be worn when plates are prepared, and all excess suspension cleaned up with a damp cloth before it solidifies. The cloth should be washed under a tap and the excess solid flushed away.

9.4.2 Applying solutions to a thin-layer plate

The technique of applying solutions to a thin-layer plate is known as 'spotting'. It is usually performed by the release of small amounts of the liquid from a capillary tube on to the surface of the layer, followed by rapid drying using a stream of air from a hair-dryer or a similar piece of equipment.

The liquid can be applied as a small circular spot but, if a fairly complex mixture is being investigated, it is better to make a short line of applications next to each other in a horizontal direction so that, after the chromatogram has been developed, short horizontal bands are formed rather than circular zones. It is useful to avoid the outer 1.5 cm from the edge of the plate when applying samples. Spotting them too low down on the plate can cause the baseline to be below the level of the solvent, and spotting them too close to the vertical edges may cause distortion in their zones due to edge effects.

9.4.3 Spraying a thin-layer plate

The most usual method of detecting colourless substances on a thin-layer plate is to spray with a chromogenic reagent. It is important that the spray is of fine-enough droplets to give a mist which wets the plate evenly; this arises from a stream of the detection fluid being forced out through a small hole under pressure.

There are several methods by which this is achieved. Aerosol spray guns are available commercially but are expensive. A good-quality spray can be obtained by means of a current of compressed air using the apparatus shown in Figure 9.4. A simple hand-held pump consisting of a

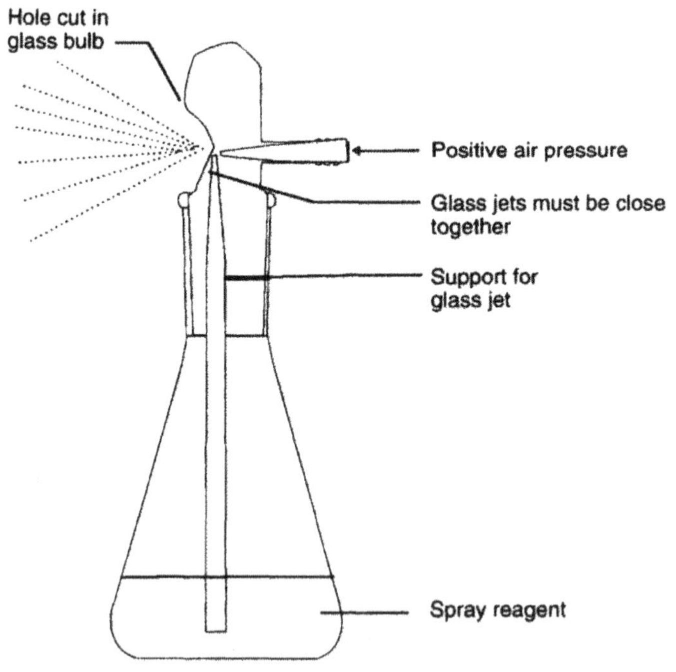

Figure 9.4 Spraying apparatus for TLC plates.

rubber bulb with a rubber reservoir can give good results and is a cheap alternative when no compressed air supply is available.

CARE!!! The plate should only be sprayed in a fume hood with a good extractor fan. Ideally excess fluid should be washed away by a constant stream of water at the back of the hood which drains to a toxic-waste collector. Eye protection is particularly important when spraying and the contamination of hands and other exposed skin should also be avoided.

The spray jet should be carefully played over the surface of the plate, sweeping in a series of horizontal lines from top to bottom and then vertical lines from left to right to ensure that all the plate is covered. It is not necessary to spray any part of the plate through which the mobile phase has not travelled.

It is important not to overwet the layer as this may cause damage to it and may also affect the shape of the zones.

9.4.4 Heating a thin-layer plate

Heating subsequent to spraying is a common part of the detection process with many chromogenic spray reagents. Sometimes the zones exhibit a sequence of colour changes which may not be the same for different compounds, although the final colour may be the same.

Heating can be carried out in a small oven, preferably with a glass door so that colour production can be observed, but special hot plates are available which give an even heating of the layer. Glass-backed plates are susceptible to cracking if placed directly on a hot plate so a mat to diffuse the heat can be used to give a more even heating.

As chemical vapours can be emitted from a plate during heating, this is best carried out in an oven fitted with an exhaust or on a hot plate placed in a fume cupboard.

9.5 BIBLIOGRAPHY

Dey, P.M. and Harborne, J.B. (series eds) *Methods in Plant Biochemistry*, Academic Press, London.

Harborne, J.B. (1984) *Phytochemical Methods*, 2nd edn, Chapman & Hall, London.

Ikan, R. (1991) *Natural Products – A Laboratory Guide*, 2nd edn, Academic Press, London.

CHAPTER TEN

Practical exercises in chemical screening and fractionation of crude extracts from plant materials

Detailed instructions are presented in this chapter for the screening of organic material for the presence of certain chemical groups (section 10.1), and exercises (section 10.2) involving specific examples of the extraction of substances from plant material, clean-up of extracts and for the separation of the individual components of extracts by chromatographic procedures.

10.1 CHEMICAL SCREENING

Chemical tests are available for use in the examination of plant extracts for the presence of a range of chemical groups (section 8.2.1). This section gives detailed instructions for the performance of these tests. In most cases, reference materials are suggested, on which the tests may be performed (in parallel with the sample) as positive controls. It must be remembered, however, that although the tests given here can give an indication of the presence of certain chemical groups, they are not sufficiently specific for this to be confirmed. The results of these tests must be taken in conjunction with those of other analytical methods before firm conclusions can be drawn. Suitable safety precautions, such as correct handling and disposal of materials and the use of operator protection (gloves, laboratory coat, safety spectacles), must always be observed in carrying out these tests.

10.1.1 Test for the presence of oils

Volatile oils have a strong smell and do not leave a permanent grease spot on paper. Fixed oils do not have a strong odour and leave a permanent translucent grease spot when applied to paper.

1. Crush a small sample of the crude drug between the thumb and fore-finger, and examine for the presence of an odour (Note 1).
2. Extract 1 g of the powdered drug by warming with 10 ml petroleum spirit (boiling point (b.p.) range 40–60°C) in a boiling tube heated on a water bath (Note 2). Do not let the solvent boil dry.
3. Filter the mixture into an evaporating dish and concentrate the filtrate to about 1 ml on a water bath. Perform the following test.
4. Using a teat pipette apply one drop of the extract to a filter paper. Expose the paper to a current of warm air and note the occurrence of any translucent area. If this is observed, then oils are present.
5. Place the paper in a oven at 105°C for 15 min. If the translucent spot can still be observed after that time, then a fixed oil is present.
6. The presence of a volatile oil is detected by the disappearance or diminution of the translucent area.

Notes:
1. Drugs containing volatile oils have a strong odour, particularly if they are crushed between the fingers.
2. Volatile oils are obtained commercially by steam distillation and not by solvent extraction.

10.1.2 Test for the presence of carotenoids

1. Warm 500 mg of the sample with 30 ml of chloroform in a test tube.
2. Filter, and divide the filtrate into two equal portions and evaporate to dryness in evaporating dishes.
3. Reconstitute each extract in 0.5 ml of chloroform.
4. To the first sample add 0.5 ml concentrated sulphuric acid. A deep blue to indigo colour indicates the possible presence of carotenoids.
5. To the second sample add 0.5 ml of a 20% w/v solution of antimony trichloride in chloroform. A permanent or transient blue colour indicates the possible presence of carotenoids.

10.1.3 Tests for the presence of alkaloids

General extraction method for alkaloids

1. Extract 2 g of powdered drug by warming for 2 min with 20 ml 1% sulphuric acid in a 50 ml conical flask on a water bath, with intermittent shaking (Note 1).
2. Centrifuge, pipette off the supernatant into a small conical flask. Make an initial test for alkaloids by adding to 0.1 ml extract in a semi-micro tube, one drop of Mayer's reagent (Appendix A); a white to buff precipitate should be formed (Note 2).
3. If a precipitate is formed, make the rest of the supernatant distinctly

alkaline with dilute ammonia solution, transfer to a 50 ml separating funnel, and extract with 3 × 10 ml dichloromethane (Note 3). Do not shake hard because the two layers will emulsify and be very difficult to separate. If done on a smaller scale, the organic and aqueous layers could be separated using a teat pipette or phase separation paper.

4. Transfer the combined organic layer to a second separating funnel and wash with a little water, i.e. shake with water but discard the aqueous layer (Notes 4 and 5).
5. Filter the dichloromethane layer (alkaloid extract) through a small plug of cotton wool placed in the neck of a filter funnel. The cotton wool removes any remaining traces of water as the dichloromethane passes through it. Pass a further 5 ml of fresh dichloromethane through the cotton wool to remove any extracted alkaloids bound to the cotton wool. This dichloromethane can be collected in the same container as the extract.
6. Divide the alkaloidal extract equally between two evaporating dishes, E1 and E2, and evaporate to dryness over a water bath in the fume cupboard. E1 will be used for general tests and a test for ergot-type indole alkaloids. E2 will be used for a test for tropane alkaloids.

Notes:
1. Alkaloids are extracted as water-soluble sulphates and the plant material is removed by centrifugation.
2. A precipitate with Mayer's reagent at this stage indicates that alkaloids may be present in the extract. A precipitate may be formed by other types of water-soluble, non-basic substances, e.g. protein or chlorophyll.
3. The addition of ammonia causes the alkaloidal bases to be released from the alkaloidal sulphates. When shaken with chloroform, the alkaloids will dissolve preferentially in the chloroform.
4. The chloroform extract is washed with a little water to remove inorganic ions.
5. The above procedure is written for the use of separating funnels (section 4.3.2). Because of unavoidable losses on a small scale, it is sometimes preferable to separate phases using specially impregnated filter paper which allows only the passage of an organic phase.

General test for the presence of alkaloids

Almost all alkaloids in acid solution give precipitates with heavy-metal reagents such as Mayer's and Dragendorff's (Appendix A) and these are known as general reagents for alkaloids.

1. Dissolve the residue from E1 in 1 ml 1% sulphuric acid.
2. Place 0.1 ml portions of the acid solution (from E1) into two semi-micro tubes.

3. To the solution in one tube, add one drop of Mayer's reagent; a white to buff precipitate is produced if alkaloids are present.
4. To the solution in the second tube, add one drop of Dragendorff's reagent; an orange-red precipitate is produced if alkaloids are present.

Test for ergot-type indole alkaloids

Reference test
To familiarize yourself with the appearance of a positive test for indole alkaloids, place 1 ml of ergotamine solution (0.2% w/v dissolved in a 65 : 35 mixture of 1% sulphuric acid and ethanol) into a semi-micro tube. Slowly add down the side of the tube about an equal volume of Ehrlich's reagent (Appendix A) so that two layers are formed. Note the colour at the junction of two layers and after shaking. A blue colour should develop.

Sample test
Transfer the remaining solution from E1 to a semi-micro tube and add Ehrlich's reagent as described above.

Semispecific test for tropane alkaloids – the Vitali–Morin test

Reference test
To familiarize yourself with a positive test for tropane alkaloids, place 1 ml of atropine solution (1% w/v in dichloromethane) in an evaporating dish, and take to dryness on a water bath. * Dissolve the residue by mixing well with two drops of fuming nitric acid (CARE!!!). Evaporate to dryness (in a fume cupboard) and moisten with 2 drops of alcoholic potassium hydroxide solution (3% w/v KOH in aldehyde-free 96% ethanol). A transient purple colour should be produced.

Sample test
Treat the residue from E2 in the same way, starting from *.

Test for the presence of xanthine alkaloids

Reference test
1. Place 1 ml of a solution containing 0.2% caffeine in chloroform into an evaporating dish.
2. Evaporate the chloroform to dryness on a water bath.
3. To the residue, add 5 drops of 30% w/v hydrogen peroxide aqueous solution and 5 drops of 2 M hydrochloric acid. Heat to dryness over a water bath.
4. On adding 1 drop of 2 M ammonia solution, the yellow-red colour of the residue should change to red-violet.

Sample test

1. Extract 500 mg of the crude drug with 10 ml of boiling water, filter while warm.
2. Cool the filtrate and extract with 2 × 5 ml chloroform.
3. Carry out the test as described for the reference from step 2 onwards.

10.1.4 Tests for the presence of glycosides

A single extract cannot be made to test for all types of glycoside. Glycosides vary widely in their chemical type, particularly in the nature of the aglycone. Their solubility properties correspondingly exhibit a wide range, and it is therefore necessary to use different extraction methods to test for different types of glycosides.

General test for the presence of glycosides in crude drugs

Test 1

1. Extract 200 mg of the sample by warming in a test tube with 5 ml of dilute (10%) sulphuric acid (test with pH paper) on a water bath at 100°C for 2 min. Centrifuge or filter; pipette off the supernatant or filtrate.
2. Neutralize the acid extract with a 5% solution of NaOH (noting the volume of NaOH added).
3. Add 0.1 ml of Fehling's solution A (Appendix A) and then Fehling's solution B (Appendix A) until alkaline (test with pH paper) and heat on the water bath for 2 min. Note the quantity of red precipitate formed and compare with that formed in test 2.

Test 2

1. Extract 200 mg of the sample using 5 ml of water instead of sulphuric acid. After boiling, add a volume of water equivalent to the volume of NaOH used in test 1 (test 1, step 2 above).
2. Add 0.1 ml of Fehling's solution A and then Fehling's solution B until alkaline (test with pH paper) and heat on the water bath for 2 min. Note the quantity of red precipitate formed (test 2).
3. Compare the quantity of precipitate formed in test 2 with that formed in test 1. If the precipitate in test 1 is greater than that in test 2, then glycosides may be present, since test 2 represents the amount of free reducing sugars already present in the crude drug, whereas test 1 represents free reducing sugars plus those released on acid hydrolysis of any glycosides in the crude drug.

Test for the presence of anthracene-derivative glycosides in crude drugs

Reference test
Carry out the tests 1 and 2 on frangula bark and compare the intensity of colour produced in the two tests. Use test tubes throughout the test.

Sample test
Repeat the two tests using the sample.

Test 1 (Borntrager test)
1. Boil 200 mg of the test material with 2 ml dilute sulphuric acid in a test tube for 5 min (Note 1).
2. Centrifuge or filter while hot (Note 2), pipette off the supernatant or filtrate, cool and shake with an equal volume of dichloromethane (Note 3).
3. Separate the lower dichloromethane layer (teat pipette) and shake with half its volume of dilute ammonia. A rose-pink to red colour is produced in the ammoniacal layer (Note 4).

Test 2 (modified Borntrager test)
1. Boil 200 mg of the test material with 2 ml dilute sulphuric acid and 2 ml 5% aqueous ferric chloride for 5 min.
2. Centrifuge or filter while hot, cool and extract with dichloromethane. Continue the test as described in test 1, step 3 (Note 5).

Notes:
1. The anthracene glycosides are hydrolysed to aglycones and sugars by boiling with acids.
2. If centrifuged hot, the plant material can be removed while these anthracene aglycones are still sufficiently soluble in hot water; however, they are insoluble in the cold.
3. When shaken with dichloromethane, the aglycones will dissolve preferentially in the dichloromethane.
4. Aglycones based on anthraquinones (which are oxidized anthracenes) give a red colour in the presence of alkali.
5. Some plants contain anthracene aglycones in a reduced form. If ferric chloride is used during the extraction, oxidation to anthraquinones takes place, which then respond to the Borntrager test. C–C Glycosides are also hydrolysed oxidatively by this procedure.

Test for the presence of cardenolide glycosides in crude drugs

Reference test
Place 2 ml of digoxin (0.1% in chloroform) solution in two evaporating dishes. Carry out tests 1 and 2 on the two samples.

Test 1 (Kedde test)
1. Evaporate the chloroform to dryness.
2. Add 1 drop of 90% alcohol and 2 drops of 2% 3,5-dinitrobenzoic acid (3,5-dinitrobenzene carboxylic acid) in 90% alcohol.
3. Make alkaline with 20% sodium hydroxide solution. A purple colour is produced (Note 1).

Test 2 (Keller–Kiliani test)
1. Evaporate the chloroform to dryness in an evaporating dish.
2. Add 0.4 ml glacial acetic acid (ethanoic acid) containing a trace of ferric chloride.
3. Transfer to a small test tube and add carefully down the side of the tube 0.5 ml of concentrated sulphuric acid. A blue colour in the upper acetic acid layer is produced (Note 2).

Sample test
1. Extract 500 mg of the sample with 6 ml of 70% alcohol by heating in a test tube on a water bath for 2 min, cool and centrifuge (Note 3).
2. Pipette off the supernatant, placing 5 ml in a thick-walled glass centrifuge tube.
3. Add 10 ml of water and 0.5 ml of a strong solution of lead subacetate (Appendix A); centrifuge, pipette off the supernatant and add 10% sulphuric acid dropwise until no further precipitate forms (Note 4).
4. Centrifuge, pipette off the supernatant and extract with two successive 5 ml portions of dichloromethane using a separating funnel (Note 5).
5. Combine the two dichloromethane extracts (lower layer), wash with 1 ml of water (Note 6), separate and filter the dichloromethane through a small plug of cotton wool placed in the neck of a filter funnel.
6. Divide the dichloromethane extract into two equal parts for tests 1 and 2 above.

Notes:
1. The colour reaction with 3,5-dinitrobenzoic acid depends upon the presence of an α,β-unsaturated-γ-lactone ring in the aglycone.
2. The 2-deoxy sugar present in these glycosides causes the blue colour.
3. 70% alcohol extracts cardiac glycosides and many other constituents, e.g. chlorophyll.
4. Lead subacetate precipitates unwanted plant pigments and 10% sulphuric acid precipitates the excess lead ions.
5. The cardiac glycosides will dissolve preferentially in the dichloromethane. Their solubility in an organic solvent despite the presence of sugars is due to the steroidal aglycone.
6. Washing with a little water removes unprecipitated lead ions which would otherwise interfere later in the test when sulphuric acid is added.

Test for the presence of saponin glycosides in crude drugs

Reference test
Carry out the following tests on saponin solution (1% in normal saline).

Test 1
Place 1 ml of the solution in a semi-micro tube. Shake and note the stable froth (Note 1).

Test 2
1. Add 0.2 ml of the solution to 0.2 ml of 10% v/v blood in normal saline and mix well.
2. Centrifuge and note the red supernatant.
3. Compare with a control tube containing 0.2 ml of 10% blood in normal saline diluted with 0.2 ml of normal saline.
4. Centrifuge and note the pale yellow supernatant and precipitate of red blood cells (Note 2).

Sample test
Extract 10 mg of the sample with 1.5 ml normal saline (Note 3) by heating in a test tube on a bath of boiling water for 2 min. Centrifuge or filter, pipette off the supernatant or filtrate. Carry out tests 1 and 2 above in semi-micro tubes.

Notes:
1. Saponin glycosides lower the surface tension of water.
2. Saponin glycosides haemolyse red blood corpuscles, releasing haemoglobin into the solution. Thus the supernatant is yellow in the control tube (normal saline does not disrupt the red blood cells) but red in the test mixture (cell membrane disrupted, so haemoglobin leaks into serum).
3. Water is normally used to extract saponins. Saline is used in this case so that the same extract can be used for test 2.

Test for the presence of cyanogenic glycosides in crude drugs

Reference test
1. Place 200 mg of wild cherry bark in a conical flask and moisten with a few drops of water. There should be no free liquid in the bottom of the flask.
2. Moisten a piece of picric acid paper with sodium carbonate solution (5% aqueous). Suspend the strip of sodium picrate (sodium 2,4,6-trinitrophenate) paper by means of a cork in the neck of the flask (Note 1).
3. Warm gently to about 37°C by placing on a lamp or thermostat-controlled water bath and allow to stand.
4. Observe the change of colour of the test paper. Hydrogen cyanide

(HCN) is liberated from cyanogenic glycosides by enzyme activity. HCN reacts with sodium picrate to form the reddish-purple sodium isopurpurate.

Sample test
Carry out an identical test on the sample in place of wild cherry bark.

Notes:
1. The test will NOT work if there is any liquid water in the flask, as the HCN produced will dissolve in the water rather than come off as a gas to react with the paper.

10.1.5 Test for the presence of phenolic compounds

Phenolic compounds may exist in the plant in free or glycosidic form.

Reference test
To 1 ml of Hamamelis extract (1 g of Hamamelis leaf extracted 20 ml of 70% ethanol) add 2 drops of 5% aqueous iron (III) chloride solution. A blue-green or green colour is produced if phenolic substances are present (Note 1).

Sample test
1. Extract 500 mg of the sample with 6 ml of 70% alcohol by heating in a test tube on a water bath for 2 min, cool and centrifuge.
2. Test the extract (Note 2) in the same way as for the reference test.

Notes:
1. Iron (III) chloride produces coloured complexes with phenolic compounds.
2. The 70% ethanol extracts phenolic substances. Most plant phenols are also easily soluble in water.

10.2 EXERCISES IN EXTRACTION, CLEAN-UP, FRACTIONATION AND CHROMATOGRAPHIC METHODS

10.2.1 List of exercises

A number of practical exercises are given in this section and are designed to give detailed instructions for some of the extraction, clean-up, fractionation and isolation methods described in earlier sections of this book. The exercises are as follows:

Section	*Exercise*
10.2.2	Extraction of hyoscine and hyoscyamine from belladonna herb using solvent extraction and solvent–solvent partition
10.2.3	Isolation of cardiac glycosides from Digitalis leaves using solvent extraction, precipitation of pigments and solvent–solvent extraction
10.2.4	Separation of glycosides of senna using exclusion chromatography
10.2.5	Use of reversed-phase silica gel to fractionate rhubarb extract by vacuum liquid chromatography
10.2.6	Fractionation of an extract of rhubarb root using flash chromatography
10.2.7	Fractionation of clove oil using vacuum liquid chromatography
10.2.8	Isolation of caffeine from tea using polyamide clean-up and solvent extraction followed by preparative thin-layer chromatography
10.2.9	Separation of alkaloids from an aqueous extract of hydrastis root using ion-exchange chromatography

10.2.2 Extraction of hyoscine and hyoscyamine from belladonna herb using solvent extraction and solvent–solvent partition

This exercise demonstrates:

1. Acid–base liquid–liquid partition procedure for extraction of alkaloids.
2. Analytical TLC of the extracted alkaloids.

Introduction

Some members of the Solanaceae, e.g. belladonna herb or root (*Atropa belladonna*), contain the important alkaloids hyoscyamine and hyoscine.

The exercise is an example of separation of alkaloids from other constituents by solvent–solvent partition using changes in pH (section 10.1.3). Any alkaloids present in the crude drug are extracted as water-soluble salts by treatment with acid. Addition of ammonia converts the alkaloids from the ionized water-soluble form to water-insoluble free bases. These can be extracted with an organic solvent such as dichloromethane. Water-soluble polar impurities present in the acid extract will not be extracted.

The extracted alkaloids can be separated easily by preparative TLC.

Materials required

- Powdered belladonna herb (species of *Datura*, *Duboisia* or *Hyoscyamus* are also suitable)

- Sulphuric acid (0.05 M)
- Conc. ammonia (13.5 M)
- Dichloromethane
- Anhydrous sodium sulphate
- Freshly prepared Dragendorff's reagent
- Fuming nitric acid
- Alcoholic potassium hydroxide solution (5%)
- Bench centrifuge
- Hot-plate
- Rotary evaporator

Confirmation of the presence of tropane alkaloids

Carry out the general test for the presence of alkaloids and the semispecific test for tropane alkaloids on a small quantity of the sample (section 10.1.3)

Preparation of a crude alkaloidal extract of belladonna herb

1. Place 20 g of belladonna herb in a dry 500 ml conical flask.
2. Add 200 ml of 0.05 M sulphuric acid. Warm the mixture on a hot-plate for 15 min, stoppering and shaking the flask occasionally.
3. Cool the mixture and divide it evenly into a number of centrifuge tubes. Centrifuge for 5 min using a bench centrifuge.
4. Filter the supernatants into a clean beaker. Add ammonia (13.5 M) until the extract is distinctly alkaline.
5. Transfer to a separating funnel and extract with 5 × 50 ml quantities of dichloromethane. If an emulsion forms, the layers can be separated by centrifugation.
6. Collect the dichloromethane (lower) layer in a conical flask, shake with anhydrous sodium sulphate to remove traces of water.
7. Take two 2 ml aliquots of the organic layer to dryness in separate evaporating dishes. Carry out the general test for alkaloids (section 10.1.3) on one sample and the semispecific test for tropane alkaloids (Vitali–Morin test, section 10.1.3) on the second to confirm that the extraction has been successful.
8. Filter the remainder of the extract in portions into a round-bottomed flask and evaporate to dryness under vacuum using a rotary evaporator.

Examination of the belladonna extract by analytical TLC

TLC system

- Stationary phase: silica gel GF_{254} (8 cm × 10 cm plates, 0.25 mm thick).
- Mobile phase: acetone : methanol : water 90 : 7 : 3.
- Reference solutions: 1% w/v hyoscine and 1% w/v hyoscyamine (in ethanol).

- Detection: UV light 254 nm before spraying. Heat for 10 min at 100°C, spray with Dragendorff's reagent. Examine in daylight for orange bands.

Method

1. Redissolve the extract in 0.5 ml chloroform.
2. Along the baseline of a small TLC plate make **two** and **six** applications of the crude extract and **two** applications of the reference solutions.
3. Develop the plate, spray and examine under daylight.

If required, the alkaloids can be separated by preparative TLC using the same TLC system.

10.2.3 Isolation of cardiac glycosides from Digitalis leaves using solvent extraction, precipitation of pigments and solvent–solvent extraction

This exercise demonstrates:

1. Extraction of cardenolides by dilute alcohol and removal of impurities by precipitation.
2. Liquid–liquid separation of cardenolides from other constituents.
3. TLC examination of cardenolides.

Introduction

Digitalis (foxglove, *Digitalis purpurea*) is one of a number of plants that contain cardioactive glycosides (cardenolides). NB In parts of the world where Digitalis is not easily obtained suitable substitutes for this exercise include common garden plants such as oleander leaves (*Nerium oleander*) and *Thevetia* species.

Cardenolides are compounds with a steroidal aglycone joined to an unsaturated lactone ring and also joined to three or four sugars (see Figure 8.2). An aqueous ethanol extract of Digitalis leaves will contain not only the cardenolides but also phenolic compounds, e.g. flavonoids, and other pigments, e.g. chlorophyll. These contaminants can be precipitated as lead complexes using lead subacetate solution, and excess lead removed as lead sulphate. The cardenolides can be extracted into chloroform. Pretreatment of the sample to remove phenolics is important, as these would otherwise also be present in the chloroform extract.

Materials required

- Dried, powdered leaves of Digitalis herb
- Ethanol (70% aqueous)
- Lead subacetate solution, strong (Appendix A)

- Sulphuric acid 10%
- Chloroform
- 2% 3,5-Dinitrobenzoic acid in 90% alcohol
- Sodium hydroxide solution (20%)
- Glacial acetic acid containing a trace of iron (III) chloride
- Concentrated sulphuric acid
- Iron (III) chloride solution (5% aqueous)
- Anhydrous sodium sulphate
- Separating funnels
- Conical flask
- Evaporating dishes
- Semi-micro test tubes
- Boiling tubes
- Round-bottomed flask
- Rotary evaporator

Confirmation of the presence of cardenolides and phenolic compounds in Digitalis herb

1. Extract 1 g of the powdered herb with 10 ml of 70% aqueous ethanol.
2. Carry out the Kedde test for cardenolides and the iron (III) chloride test for phenolic compounds as described in sections 10.1.4 and 10.1.5.

Initial extraction and removal of phenolics by precipitation

1. Extract 20 g of the powdered herb with 100 ml of ethanol 70% by warming on a hot-plate for 20 min.
2. Cool and filter or centrifuge to remove solid plant debris.
3. To the filtrate or supernatant add 150 ml of water and 20 ml of a strong solution of lead subacetate; the phenolic compounds should precipitate as insoluble lead complexes.
4. Centrifuge the mixture, pipette off the supernatant and add 10% sulphuric acid dropwise until no further precipitate forms. This treatment removes excess lead ions as insoluble lead sulphate.

Solvent partition to remove cardenolides from the aqueous layer

1. Centrifuge the mixture, pipette off the supernatant and extract with four successive 50 ml portions of chloroform.
2. Combine the chloroform extracts, wash with 20 ml of water to remove any residual lead ions.
3. Separate the chloroform extract and dry using anhydrous sodium sulphate and filter the chloroform into a round-bottomed flask.

4. Evaporate three 5 ml aliquots of the chloroform to dryness in separate evaporating dishes and carry out the Kedde test for cardenolides and the iron (III) chloride test for phenolic compounds on the residues (sections 10.1.4. and 10.1.5).

Examination of the cardenolides in the crude extract using analytical TLC

Materials
- Three large TLC tanks with lining paper.
- One small TLC tank with lining paper.
- Silica gel G plates (8 × 10 cm) 0.25 mm layer.
- Mobile phase: ethylacetate : methanol : water 81 : 11 : 8.
- Detection reagent: Kedde's reagent A (3,5-dinitrobenzoic acid (3%) in ethanol), followed by Kedde's reagent B (aqueous sodium hydroxide 2 M).
- Digoxin reference solution 0.5% w/v in ethanol.

Method
1. Pour mobile phase into the small TLC tank to a depth of about 0.5 cm, close lid of tank and leave to equilibrate.
2. Reconstitute the residue of crude cardenolide extract from the first part of the experiment with about 1 ml ethanol.
3. To a small (8 cm × 10 cm) TLC plate, apply along the baseline (1.5 cm from the edge of one of the 8 cm sides) separate short bands 5 mm wide of 10 and 20 µl of the extract, interspersed with 5 and 10 µl applications of the reference solution.
4. When the spots are dry, place the plate in the tank and allow the solvent to rise to a height of 8 cm from the baseline.
5. Remove the plate from the tank and dry in a fume cupboard, using a hair-dryer to speed up the process if necessary.
6. Spray the plate with the detection reagents. Cardenolides will stain violet with this agent.
7. Note the position of the spots and confirm the presence of cardiac glycosides in the extract.

10.2.4 Separation of glycosides of senna using exclusion chromatography

This exercise demonstrates:

1. Extraction of anthracene-derivative glycosides by solvent extraction.
2. Separation of glycosides according to size by exclusion chromatography.
3. TLC examination of fractions obtained from exclusion chromatography.

Introduction

The leaves and fruits of senna *Cassia acutifolia, C. angustifolia* are the preferred treatment for constipation in many clinical circumstances. These plants contain a complex mixture of anthracene derivatives which vary considerably in molecular size. In order of increasing size these are the free anthraquinones, the anthraquinone glycosides, the dianthrone aglycones and the dianthrone glycosides.

The differences in size enable these types to be separated by exclusion chromatography. Since the non-glycosides are not very soluble in water, an alcoholic extract is used and separation effected on Sephadex LH-20® gel, a partially methylated and therefore less polar stationary phase, which swells in dilute alcohol.

The compounds are eluted in order of decreasing molecular size and this is monitored by TLC of the fractions from the column.

Materials required

- Powdered senna leaves (similar crude drugs such as frangula, cascara and rhubarb can be used)
- 70% Ethanol
- Reflux condenser
- Water bath
- Buchner funnel + flask
- Rotary evaporator
- Sephadex LH-20
- Column, reservoir, etc. for exclusion chromatography
- Fraction collector, test tubes, etc.
- Spotting capillaries
- Silica gel GF_{254} plates, 0.25 mm thick, 20 × 20 cm
- Propan-l-ol : ethyl acetate : water 40 : 40 : 30
- Spray reagents: nitric acid and 5% w/v ethanolic KOH

Preparation of the extract

1. Extract 5 g of coarsely powdered senna leaves with 30 ml of 70% ethanol by refluxing on a water bath for 15 min.
2. Filter under reduced pressure and take the filtrate to dryness using reduced pressure.
3. Dissolve the residue in 3 ml 70% ethanol and retain 0.5 ml of this as a reference solution for TLC examination. Filter the extract solution if necessary to remove any solid particles.

Preparation of the column

1. Weigh out 20 g Sephadex LH-20® gel and allow to swell for 24 h in 100 ml 70% ethanol.
2. Pour the slurry into the column with the column outlet closed and ensure that the column is full.
3. Make the appropriate connections to the column from the reservoir and from the column to the fraction collector.
4. Fill the reservoir with 70% ethanol, open the column tap and allow the ethanol to flow through the column for about 30 min. Use the outlet tap to adjust the flow to about 1 ml/min.
5. Introduce 1.0 ml of the extract on to the top of the column and let the mobile phase flow through the column. Collect eluant fractions at 2 min intervals.
6. Concentrate each eluate to about 0.5 ml under reduced pressure. Examine the eluants by TLC using the total extract as a reference.

Examination of the eluates by analytical TLC

TLC system
- Stationary phase: silica gel GF_{254} plates, 0.25 mm thick, 20 × 20 cm.
- Mobile phase: propan-l-ol : ethyl acetate : water 40 : 40 : 30
- Detection: UV light 254 and 365 nm before spraying. Daylight after spraying with nitric acid (CARE!!! This is corrosive), heating at 105°C for 15 min, cooling and overspraying with 5% ethanolic KOH.

Method
1. Apply 5 µl of the total extract as a short band in the centre of the baseline on the TLC plate.
2. Apply 5 µl aliquots of each eluate as short bands along the baseline, on either side of the total extract band.
3. Develop the plate in the mobile phase, and examine under UV light and after spraying.

10.2.5 Use of reversed-phase silica gel to fractionate rhubarb extract by vacuum liquid chromatography

This exercise demonstrates:

1. Extraction of anthracene derivatives and polyphenols with water.
2. Separation of polar from less polar constituents by the use of reverse-phase sorbent extraction.

Introduction

Rhubarb root, *Rheum palmatum*, contains a variety of water-soluble compounds, including polyphenolic tannins, anthracene glycosides and free anthraquinones. All of these contribute to the biological activity of rhubarb which is used pharmaceutically both as a treatment for diarrhoea (due to the tannins) and as a laxative (the anthraquinone compounds).

The polarity of these three groups of compounds is different – the tannins being most polar and the free anthraquinones least polar. These differences in polarity are exploited in the separation of these groups from each other by reverse-phase chromatography. The stationary phase is less polar relative to the mobile phase so the more polar compounds, i.e. the tannins, should be eluted first. Elution is carried out by adding aliquots of mobile phase of decreasing polarity.

Reverse-phase silica acts as a non-polar adsorbent since it has a non-polar octadecyl coating. If reverse-phase silica is unavailable, thoroughly cleaned and washed powdered charcoal can be used as a substitute. If charcoal is used alone, it tends to clog up the pores in the filter paper so a kieselguhr bed should be made to aid filtration (section 9.1.1).

Materials needed for reverse-phase bed fractionation

- Powdered rhubarb root
- Water
- Dichloromethane
- Test tubes
- Dilute sulphuric acid
- Dilute ammonia
- Ferric chloride (5%, aqueous)
- Fehling's solution A and B
- Litmus paper
- Sintered glass funnel and flask (or T-piece adaptor + separating funnel for connection)
- Reverse-phase silica
- Water
- Methanol
- Acetone
- Silica gel TLC plates
- 5% Aqueous iron (III) chloride spray reagent
- 5% KOH in ethanol spray reagent
- TLC tanks
- Developing solvent for TLC: ethyl acetate : methanol : water 100 : 13.5 : 10
- Rotary evaporator
- Round-bottom flasks

Confirmation of the presence of anthracene-derivative glycosides in rhubarb root

Carry out the general test for the presence of glycosides and the semi-specific tests for anthracene-derivative glycosides (section 10.1.4) on a small sample of the rhubarb powder.

Preparation of extract

1. Weigh out 10 g of powdered rhubarb and heat with 100 ml water for 30 min.
2. Connect up the Buchner filter and funnel to the vacuum line.
3. Moisten the filter paper and introduce it into the Buchner funnel making sure that there are no leaks. Keep the vacuum line closed.
4. Prepare a slurry of kieselguhr in water and pour it into the Buchner filter.
5. Add 20 g kieselguhr to the rhubarb and water suspension and filter through the kieselguhr bed on the Buchner filter. Concentrate 5 ml of the filtrate to serve as a reference solution for TLC analysis of the eluates.

Preparation of reverse-phase silica bed (alternative method to charcoal)

1. Connect up the Buchner filter and funnel to the vacuum line.
2. Prepare a slurry of 30 g reverse-phase (C18 or C8) silica in methanol and pour it into the Buchner filter. Keep the vacuum on but only slightly.
3. Wash the bed with 100 ml water to elute off all the methanol (traces of methanol will diminish the adsorption of non-polar components on the silica).
4. Do not let the bed dry out – add more water as necessary.

Preparation of charcoal bed (alternative to a reverse-phase silica bed)

1. Connect up the Buchner filter and funnel to the vacuum line.
2. Moisten the filter paper and introduce it into the Buchner funnel, making sure that there are no leaks. Keep the vacuum line closed.
3. Prepare a slurry of kieselguhr in water and pour it into the Buchner filter.
4. Weigh out approximately equal quantities of charcoal and kieselguhr sufficient to fill about half of the filter funnel. Mix together with water to form a slurry.
5. Pour the slurry on to the kieselguhr bed. Keep the vacuum on but only slightly.
6. Do not let the bed dry out – add more water as necessary.

Fractionation of extract

1. Carefully pour the extract on to the silica (or charcoal) bed and elute with two bed volumes of water.
2. When the water has passed through, allow air to pass through the reversed-phase silica bed then release the vacuum.
3. Transfer the eluate to a suitable flask and concentrate it under reduced pressure.
4. Close the vacuum line and elute the bed with two bed volumes of water : methanol 1 : 1.
5. Collect the eluate and transfer as in step 3.
6. Repeat elutions as in step 4 using successively methanol, methanol : acetone 1 : 1 and acetone. Concentrate 5 ml of each eluate to low volume under reduced pressure.

Examination of the eluates by analytical TLC

TLC system
- Silica gel G plates, 0.25 mm thick, 20 × 20 cm.
- Mobile phase: ethyl acetate : methanol : water 100 : 13.5 : 10.

Method
Examine the eluates by TLC using the total extract as a reference. Use two plates for each fraction. After development of the TLC plates examine by spraying one plate with the iron (III) chloride solution (to detect tannins) and one with the 5% ethanolic KOH (to detect anthraquinones). Note the separation of the different constituents.

10.2.6 Fractionation of an extract of rhubarb root using flash chromatography

This exercise demonstrates:

1. Extraction of anthracene-derivative glycosides by alcohol.
2. Gradient elution flash column chromatography.

Introduction

The roots of rhubarb (*Rheum* spp.) have been widely used in Chinese and European medicine because of their anthracene-derivative glycosides and also for their content of polyphenols (also known as tannins) (section 10.2.5).

Materials required

- Flash column apparatus
- Silica gel for flash chromatography

- Ethyl acetate
- Methanol
- Water
- Nitrogen cylinder and lead
- Fraction collector with test tubes
- Silica gel TLC plates
- 5% Aqueous iron (III) chloride spray reagent
- 5% KOH in ethanol spray reagent
- TLC tanks
- Developing solvent for TLC: ethyl acetate : methanol : water 100 : 13.5 : 10
- Rotary evaporator
- Round-bottom flasks

Preparation of the extract

1. Extract 5 g coarsely powdered rhubarb root with 30 ml of 70% ethanol by refluxing on a water bath for 15 min.
2. Filter under reduced pressure and take the filtrate to low volume using reduced pressure.
3. Mix the residue with a small amount of the same type of silica gel as used in the column.
4. Stir well and allow to dry overnight.

Preparation of column

1. Measure out enough silica gel to fill about half of the column. Tip the silica into a beaker and mix it with ethyl acetate to form a loose slurry.
2. Ensure that the tap at the bottom of the flash column is closed then pour the slurry into the column. Allow to settle and wash down any slurry adhering to the walls of the column with small amounts of ethyl acetate.
3. Open the tap on the bottom of the column and place a flask under the elution tube to collect waste ethyl acetate.
4. Attach the nitrogen line to the top of the column and carefully turn on the gas until about 5 bar is reached. Ensure that all taps are shut to minimize gas leakage.
5. BEFORE the top of the ethyl acetate reaches the top of the silica, open the bleed valve on the gas lead, close the control on the cylinder head and close the tap at the base of the column. There should now be less than 1 cm of mobile phase above the top of the stationary phase in the column.

Separation of the components of the extract

1. Disconnect the top of the column and add the extract adsorbed on silica to the small amount of ethyl acetate above the top of the silica in the column.
2. Open the tap at the bottom of the column, connect the gas supply and open carefully so that the liquid is level with the top of the silica. Close the bottom tap.
3. Disconnect the top of the column, fill the column with ethyl acetate, reconnect the gas supply, open the tap at the bottom of the column, turn on the gas and start collecting fractions.
4. When the eluant is near the top of the layer, release the gas pressure, close the tap at the bottom of the column, disconnect the top of the column and fill the column with more ethyl acetate.
5. Reconnect the top of the column and gas supply. Open the tap at the bottom of the column and collect more fractions.
6. Repeat steps 4 and 5 using two amounts of ethyl acetate : methanol 9 : 1 then amounts of ethyl acetate : methanol : water 90 : 5 : 5.

Examination of eluates by analytical TLC

TLC system
- Stationary phase: silica gel G plates, 0.25 mm thick, 20 × 20 cm.
- Mobile phase: ethyl acetate : methanol : water 100 : 13.5 : 10.
- Detection method: examine under UV light at 254 nm. Spray with 5% ethanolic KOH and examine in daylight.

Method
Use TLC to examine the fractions obtained and use the total extract as a reference on each TLC plate used. It is a good idea to investigate initially only a representative number of the fractions, e.g. every fifth fraction.

10.2.7 Fractionation of clove oil using vacuum liquid chromatography

This exercise demonstrates:

1. Separation of the constituents of clove oil by normal phase vacuum liquid chromatography (VLC).
2. TLC examination of fractions obtained.

Introduction

Clove oil is used as a flavouring agent, stimulant and antiseptic, particularly in dentistry. It owes its antiseptic properties to the presence of phenolic substances, particularly eugenol. The major two types of constituents in clove oil, i.e. the non-polar hydrocarbon terpenes and the comparatively polar phenolic eugenol, can be separated by vacuum liquid chromatography using silica gel. Elution is carried out in order of increasing polarity, starting with a very non-polar solvent. Thus the hydrocarbons are first eluted and the eugenol is retained on the bed until it is eluted with a more polar solvent. The fractions are monitored using TLC.

Materials required

- Oil of clove (NB This can be obtained from cloves by steam distillation (section 3.3.2), allow 30 g cloves for 1 ml oil)
- Sintered glass or Buchner filter + flask (or T-piece adaptor and separating funnel)
- TLC-grade silica gel
- Round-bottom flasks
- Rotary evaporator
- Silica gel G TLC plates, 0.25 mm thick, 20 × 20 cm
- Developing solvent: toluene
- Detection spray reagent: 0.5% v/v anisaldehyde in methanol : acetic acid : sulphuric acid 85 : 10 : 5
- TLC tanks

Preparation of the sample for VLC

1. Dissolve about 0.25 g of the clove oil in a few millilitres of light petroleum (b.p. range 40–60°C). Mix it with a small amount of TLC-grade silica gel in an evaporating basin until a paste is formed.
2. Stir occasionally with a glass rod to aid evaporation of the solvent. The material must be completely dry and in the form of a free-flowing powder before use.

Preparation of the VLC bed (column)

1. Assemble the sintered glass or Buchner filter and collection apparatus.
2. Attach to a vacuum line and ensure that the system is under vacuum and that the vacuum can be controlled if necessary to give only slight negative pressure.
3. Use sufficient TLC-grade silica gel to fill not more than half the funnel and use it to make a slurry with light petroleum (b.p. range 40–60°C) (hexane can also be used).
4. Pour the slurry on to the filter bed. Apply slight negative pressure so that the light petroleum is sucked through the bed but release the vacuum before the bed goes dry.
5. Transfer the collected solvent to a suitable vessel. Reconnect the apparatus.

Fractionation of the clove oil sample by VLC

1. When the sample paste has completely dried and is a free-flowing powder, carefully apply it to the silica bed, ensuring that the bed is evenly covered.
2. Add more light petroleum to the top of the bed and apply negative pressure until all the oil solution is just absorbed into the silica. Release the vacuum.
3. Apply enough light petroleum (b.p. range 40–60°C) to fill the rest of the funnel and then apply slight negative pressure to the system. Collect the eluate.
4. Repeat step 2 with a second aliquot of light petroleum (b.p. range 40–60°C).
5. Concentrate the collected eluates using the rotary evaporator.
6. Repeat steps 2–4 using successive 100 ml aliquots of light petroleum (b.p. range 40–60°C) : chloroform 1 : 1, chloroform, chloroform : ethyl acetate 1 : 1, ethyl acetate.
7. Concentrate the eluates to a small volume using the rotary evaporator.

Analysis of VLC eluates using analytical TLC

TLC system

- Stationary phase: silica gel G TLC plates, 0.25 mm thick, 20 × 20 cm.
- Mobile phase: toluene.
- Detection reagent: spray with 0.5% v/v anisaldehyde in methanol : acetic acid : sulphuric acid 85 : 10 : 5.
- Reference solution: 0.5% eugenol in chloroform.
- Visualization: daylight after heating for 10 min at 105°C after spraying.

Method

1. Dilute a sample of clove oil to 1% v/v in chloroform.
2. Apply 5 µl of the oil solution as a short band to the centre of the baseline of the TLC plate.
3. Apply 5 µl of the eugenol solution as a short band next to the clove oil band.
4. Apply 5 µl aliquots of each of the concentrated VLC eluates as short bands on either side of the reference (oil and eugenol) solutions.
5. Develop the plate, spray it and heat for the required time. Then examine in daylight.

The first fractions to elute should contain compounds with high R_f values which give a colour after spraying. The fractions containing eugenol should elute only with the chloroform or more polar eluants.

10.2.8 Isolation of caffeine from tea using polyamide clean-up and solvent extraction followed by preparative thin-layer chromatography

This exercise demonstrates:

1. Removal of polyphenols in an aqueous extract by complexation with polyvinylpolypyrrolidone (PVPP).
2. Partition system for separation of caffeine from other water-soluble compounds.
3. Analytical and preparative TLC to isolate caffeine.

Introduction

The alkaloid caffeine belongs to a group of compounds, known as xanthines, which are based on a purine nucleus (Figure 8.1). Related alkaloids include theobromine and theophylline. The compounds differ in the position and number of methyl substituents in the purine ring. They are variously distributed in common beverages such as coffee, tea and cocoa.

Commercial black tea leaves (*Camellia sinensis*) contain caffeine (2.5%), theobromine (0.17%) and theophylline (0.013%). They also contain considerable amounts of oxidized polyphenols (tannins) which account for the dark colour of an infusion of the leaves. In this experiment, the xanthine alkaloids are extracted using hot water, and the polyphenols removed by passing the extract through a column of polyvinylpolypyrrolidone (PVPP). Phenols bind to the polyamide groups of this material, whereas the xanthines elute.

Caffeine can be extracted from the aqueous layer into chloroform. Theobromine is only poorly soluble in chloroform and will remain in the aqueous layer. Theophylline will be extracted into chloroform but may not be detected due to its low concentration. (The polyphenols in tea may be extracted into chloroform if they are not removed in the earlier step.)

Materials

- Black tea leaves (powdered coffee beans, cocoa, mate tea or guarana are suitable substitutes)
- Boiling water
- Polyvinylpolypyrrolidone (PVPP, chromatography grade)
- Chloroform
- Sodium sulphate (anhydrous)
- Ammonia (2 M)
- Beaker
- Separating funnels (50 ml and 500 ml)
- Syringe (20 ml) with adaptor for vacuum manifold
- Manifold for sorbent extraction
- Collection flask

- Round-bottomed flask
- Rotary evaporator
- Conical flask
- One small TLC tank with lining paper
- Silica gel GF_{254} plates, 8 × 10 cm, 0.25 mm (analytical)
- Silica gel PF_{254} plates, 20 × 20 cm, 0.5 mm thick layer (preparative)
- Caffeine reference solution, 0.5% w/v in ethanol
- Theophylline reference solution, 0.5% w/v in ethanol
- Theobromine reference solution, 0.5% w/v in ethanol

Hot-water extraction of caffeine from tea leaves

1. Place about 10 g of tea leaves in a beaker and add 200 ml of boiling water. Heat on a hot-plate for about 5 min.
2. Filter while hot to remove the tea leaves and allow to cool. Carry out a test for phenolic compounds on the solution (section 10.1.5).

Removal of polyphenols from the extract

1. Stir 10 g polyvinylpolypyrrolidone (PVPP) into the tea extract and leave to settle.
2. Decant off the supernatant and perform the test for phenolic compounds on 0.4 ml of the supernatant.
3. If a positive result is obtained, repeat the procedure with a further 10 g of PVPP.

Concentration of the caffeine extract

1. Place the cleaned water extract in a separating funnel and extract with 4 × 50 ml chloroform.
2. Combine the chloroform extracts in a conical flask and shake with sodium sulphate (anhydrous) to remove any remaining traces of water.
3. Filter the chloroform solution, in appropriate portions, into a round-bottomed flask and evaporate to dryness (preferably under vacuum using a rotary evaporator).

Confirmation of the presence of caffeine in the extract by analytical TLC

TLC system

- Stationary phase: silica gel GF_{254}.
- Mobile phase: chloroform : ethanol 95 : 5.
- Detection reagent: iodine-hydrochloride reagent. A: potassium iodide (1 g) and iodine (1 g) are dissolved in ethanol (100 ml). B: hydrochloric acid (25%, 25 ml) is mixed with ethanol (96%, 25 ml). Spray with reagent A, followed by reagent B. Examine in daylight.

Separation of xanthine alkaloids in the crude extract on preparative TLC plates

Use the same mobile phase as for the analytical TLC but use 20 cm × 20 cm plates at least 0.5 mm thick, and made of silica gel PF_{254}.

1. Starting 2 cm from the bottom of the plate and 3.5 cm from the left-hand edge, apply the extract as a band 13 cm long along the baseline. Do this by making individual applications side by side.
2. When the entire line has been applied once and is dry, repeat the process until about 0.25 ml of extract has been applied, ensuring that the line is dry between applications (otherwise, the sample will be applied as a diffuse band and consequently poor resolution of components will occur). Take care not to damage the silica layer of the plate during repeated applications.
3. Apply 5 µl of the caffeine reference solution as a spot on the baseline 2 cm from the right-hand edge of the plate. Repeat 2 cm from the left-hand edge.
4. Place each plate in a separate tank and allow the solvent to rise to a height of 15 cm from the baseline.
5. Remove the plate from the tank and allow to dry **in the fume cupboard**. Use a hair-dryer if necessary.
6. Examine the plate under UV light at 254 nm and mark the position of any spots and bands using a mounted needle.
7. Based on the position of the reference spots, identify the band corresponding to caffeine in the sample.

Removal of the caffeine band from the preparative TLC plate

1. Wearing a **face mask** to prevent inhalation of silica dust, scrape off the silica corresponding to the band on to a piece of paper. Take care not to include any silica from outside the boundary, or from the sprayed area.
2. Transfer the silica samples from the three plates to a 200 ml conical flask and add 100 ml acetone. Shake to ensure that the silica gel is mixed well with the acetone and allow to stand for 5 min. At the end of this time filter the acetone into a round-bottomed flask.
3. Evaporate the acetone to a small volume under reduced pressure (using a rotary evaporator).
4. Transfer the solution to a preweighed evaporating dish, along with acetone washings of the round-bottomed flask (small volumes), and complete the evaporation over a water bath.
5. Weigh the evaporating dish and calculate the weight of the residue.

Examination of the isolated caffeine by analytical TLC

Repeat the analytical TLC to confirm that the caffeine has been isolated.

1. Dissolve a small quantity of the residue in about 100 µl of chloroform.
2. Apply 5 and 10 µl aliquots of this solution, the original crude extract and the reference caffeine solution as adjacent 5 mm bands on a 8 cm × 10 cm plate.
3. Develop with mobile phase, dry and spray as before.
4. The success of the preparative method in isolating caffeine can now be evaluated.

10.2.9 Separation of alkaloids from an aqueous extract of hydrastis root using ion-exchange chromatography

This exercise demonstrates:

1. Extraction of quaternary and tertiary amine alkaloids by acidic buffer.
2. Separation of alkaloids from other constituents by ion-exchange chromatography.
3. Precipitation of alkaloids by complexation with Mayer's reagent and subsequent regeneration of the free base form by ion exchange.

Introduction

Hydrastis root, *Hydrastis canadensis,* contains isoquinoline alkaloids, including the quaternary amine berberine which has a bright yellow colour. Berberine has a variety of biological activities and is used in some countries as a treatment for the protozoal disease leishmaniasis.

The alkaloids are extracted with a buffer of low pH and all thus acquire a positive charge by protonation of the N atom. The quaternary alkaloids exist naturally in a charged form. Two different uses of ion-exchange resins can be applied to separate the alkaloids from other components present in the aqueous extract. One method uses ion-exchange resin to convert alkaloid–heavy metal complexes to alkaloid chlorides, while the other method uses a negatively charged resin to retain the alkaloidal cations which are then subsequently eluted.

In method 1 the alkaloids are precipitated as complexes with Mayer's reagent (potassium mercuric iodide) and are thus separated from the other water-soluble constituents. The complex is not soluble in water but is soluble in less-polar solvents. The mercuric iodide portion of the complex has an overall negative charge and ion-exchange is used to replace it with Cl^- to form the soluble chlorides of the alkaloids. The mercury complex is retained by the cationic resin used and the alkaloids are eluted from the column as the chlorides. The eluate containing alkaloids can be concentrated and other preparative methods used to isolate individual alkaloids.

Method 2 uses the anionic resin Amberlite CG-120 to retain the alkaloids from the acidic extract of the plant. The weaker bases are then eluted with water and the stronger bases with 2 M ammonia solution. At the pH of water the weaker bases lose their charge and thus their affinity for the resin. The stronger bases, particularly the quaternary alkaloids, retain their charge even at high pH and elution occurs due to the displacement of the alkaloid cations by the ammonium cations, which have a higher affinity for the anionic ligands on the ion-exchange resin.

Materials required

- Powdered hydrastis root (NB The stems and roots of some other plants, including the garden ornamentals such as *Berberis* and *Mahonia* species, contain the same type of compounds and can be used as substitutes).

Extraction
- Buffer pH 2.8
- Buchner funnel and flask
- Filter paper

Clean-up method 1
- Mayer's reagent (Appendix A)
- Chromatography column
- IRA 400 Cl^- ion-exchange resin
- Acetone : methanol : water 6 : 2 : 1
- Rotary evaporator
- Silica gel GF_{254} TLC plates
- Propan-l-ol : formic acid : water 90 : 1 : 9
- Dragendorff's spray reagent
- UV light
- Berberine reference solution
- Small flasks

Clean-up method 2
- Chromatography column
- Amberlite CG-120 ion-exchange resin
- 2 M ammonia
- Water
- Rotary evaporator
- Silica gel GF_{254} TLC plates
- Propan-l-ol : formic acid : water 90 : 1 : 9
- Dragendorff's spray reagent
- UV light
- Berberine reference solution
- Small flasks

Extraction of the alkaloids

1. Weigh out 20 g of the powdered hydrastis and heat with 40 ml of the buffer, pH 2.8, for 15 min.
2. Filter through a Buchner filter. Retain 1 ml of the filtrate to serve as a reference solution for TLC analysis of the eluates.
3. Divide the rest of the filtrate into two equal portions for clean-up methods 1 and 2.

Clean-up method 1: precipitation with Mayer's reagent

Preparation of the resin
1. Weigh out 25 g of the IRA 400 resin and wash it with 5% aqueous NaOH, water and dilute HCl.
2. Wash the resin thoroughly with water and then suspend it in acetone : methanol : water 6 : 2 : 1 and pour into a column with the drain tap closed.
3. Allow to stand for at least 2 hours.

Treatment of plant extract
1. Add Mayer's reagent to the extract until no more precipitate forms.
2. Filter the suspension using a Buchner funnel. Wash the residue on the filter bed with water.
3. Discard the filtrate and dissolve the residue in acetone : methanol : water 6 : 2 : 1 by allowing the solvent to percolate slowly, i.e. no applied vacuum, through the filter bed.

Clean-up using ion-exchange
1. Open the tap at the bottom of the column and allow the suspending fluid to run out. Concentrate this eluant.
2. Apply the solution of the Mayer's precipitate in acetone : methanol : water 6 : 2 : 1 to the top of the column and allow it to pass through the column under gravity. Collect 10 ml aliquots of the eluant.
3. When all the solution has been applied, wash the column with more of the acetone : methanol : water mixture until the eluant does not give a positive alkaloid reaction when spotted on filter paper and sprayed with Dragendorff's reagent.

Examination of eluates from clean-up method 1 using analytical TLC
Concentrate each eluate and examine by TLC using the original extract as a reference.

TLC system

- Stationary phase: silica gel GF_{254} plates, 0.25 mm thick 20 × 20 cm.
- Mobile phase: propan-l-ol : formic acid : water 90 : 1 : 9.
- Detection: under UV light 254 nm and 365 nm; daylight after spraying with Dragendorff's reagent.

Clean-up method 2

Preparation of the resin

1. Weigh out 25 g of the Amberlite CG-120 resin and wash it with 2 M ammonia, water and buffer pH 2.8.
2. Suspend the resin in buffer pH 2.8 and pour into a column with the drain tap closed. Allow to stand for at least 2 hours.

Separation using ion-exchange

1. Open the tap at the bottom of the column and allow the buffer to run out.
2. Apply the plant extract in buffer pH 2.8 to the top of the column and allow it to pass through the column under gravity. Collect 10 ml aliquots of the eluant.
3. When all the extract has been applied, wash the column with another 30 ml of the buffer.
4. Elute the column with 40 ml water, collecting 10 ml eluant aliquots as before.
5. Elute the column with 40 ml 2 M ammonia, collecting 10 ml eluant aliquots as before.

Examination of eluates using analytical TLC

Concentrate each eluate and examine by TLC as above using the original extract as a reference.

Recipes for chemical test reagents

DRAGENDORFF'S REAGENT

- Stock solution: mix bismuth subnitrate (oxynitrate; 1.7 g) with water (80 ml) and glacial acetic acid (20 ml). Add potassium iodide solution (50% w/v, 100 ml). Shake or stir until dissolved. Solution keeps indefinitely when stored in a dark bottle.
- Working solution: mix the stock solution (100 ml) with glacial acetic acid (200 ml) and make up to volume (1 litre) with distilled water. Keeps for 2–5 months when stored in a dark bottle.

EHRLICH'S REAGENT

Add *p*-dimethylaminobenzaldehyde (0.25 g) to concentrated sulphuric acid (130 ml) cooled in ice. Add this slowly to water (70 ml) and then add iron (III) chloride solution (5% aqueous; 0.2 ml). The solution is pale yellow when fresh, and dark yellow when decomposed.

FEHLING'S REAGENT A

Dissolve copper sulphate (70 g) and concentrated sulphuric acid (1 ml) in distilled water (200 ml) and make up to volume (1 litre) with distilled water.

FEHLING'S REAGENT B

Dissolve sodium potassium tartrate (352 g) and sodium hydroxide (154 g) in distilled water (200 ml) and make up to volume (1 litre) with distilled water.

KEDDE REAGENT A

Dissolve 3,5-dinitrobenzoic acid (2 g) in 90% ethanol (100 ml).

KEDDE REAGENT B

Dissolve sodium hydroxide (5 g) in distilled water (100 ml).

LEAD SUBACETATE, STRONG

Dissolve lead (II) acetate (40.0 g) in carbon dioxide-free water (90 ml). Adjust the pH to 7.5 with 10 M sodium hydroxide aqueous solution. Centrifuge and use the clear supernatant. Store in a well-closed container.

MAYER'S REAGENT

Dissolve mercuric chloride (2.72 g) in distilled water (120 ml). Separately dissolve potassium iodide (10 g) in distilled water (40 ml). Mix the two solutions and make up to volume (200 ml) with distilled water.

Index

GPSR Compliance
The European Union's (EU) General Product Safety Regulation (GPSR) is a set of rules that requires consumer products to be safe and our obligations to ensure this.

If you have any concerns about our products, you can contact us on

ProductSafety@springernature.com

In case Publisher is established outside the EU, the EU authorized representative is:

Springer Nature Customer Service Center GmbH
Europaplatz 3
69115 Heidelberg, Germany

www.ingramcontent.com/pod-product-compliance
Ingram Content Group UK Ltd.
Pitfield, Milton Keynes, MK11 3LW, UK
UKHW010216160926
475607UK00001B/5
* 9 7 8 0 4 1 2 7 4 9 1 0 0 *